Trends
in
Chemical Consulting

Trends in Chemical Consulting

Charles S. Sodano, EDITOR
Nabisco Biscuit Company

David M. Sturmer, EDITOR
Eastman Kodak Company

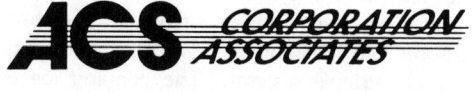

ACS CORPORATION ASSOCIATES
Annual Symposium

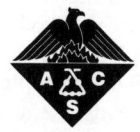

American Chemical Society, Washington, DC 1991

Library of Congress Cataloging-in-Publication Data

Trends in chemical consulting / Charles Sodano, David Sturmer.

    p.   cm.

Includes index.

ISBN 0-8412-2106-5

1. Chemical engineering consultants—Vocational guidance.

I. Sodano, Charles S.   II. Sturmer, David.   III. American Chemical Society.

TP186.T74   1991
660—dc20   91-20718
                                      CIP

Copyright © 1991

American Chemical Society

All Rights Reserved. The appearance of the code at the bottom of the first page of each chapter in this volume indicates the copyright owner's consent that reprographic copies of the chapter may be made for personal or internal use or for the personal or internal use of specific clients. This consent is given on the condition, however, that the copier pay the stated per-copy fee through the Copyright Clearance Center, Inc., 27 Congress Street, Salem, MA 01970, for copying beyond that permitted by Sections 107 or 108 of the U.S. Copyright Law. This consent does not extend to copying or transmission by any means—graphic or electronic—for any other purpose, such as for general distribution, for advertising or promotional purposes, for creating a new collective work, for resale, or for information storage and retrieval systems. The copying fee for each chapter is indicated in the code at the bottom of the first page of the chapter.

The citation of trade names and/or names of manufacturers in this publication is not to be construed as an endorsement or as approval by ACS of the commercial products or services referenced herein; nor should the mere reference herein to any drawing, specification, chemical process, or other data be regarded as a license or as a conveyance of any right or permission to the holder, reader, or any other person or corporation, to manufacture, reproduce, use, or sell any patented invention or copyrighted work that may in any way be related thereto. Registered names, trademarks, etc., used in this publication, even without specific indication thereof, are not to be considered unprotected by law.

PRINTED IN THE UNITED STATES OF AMERICA

1991 ACS Books Advisory Board

V. Dean Adams
Tennessee Technological
  University

Paul S. Anderson
Merck Sharp & Dohme
  Research Laboratories

Alexis T. Bell
University of California—Berkeley

Malcolm H. Chisholm
Indiana University

Natalie Foster
Lehigh University

Dennis W. Hess
University of California—Berkeley

Mary A. Kaiser
E. I. du Pont de Nemours and
  Company

Gretchen S. Kohl
Dow-Corning Corporation

Michael R. Ladisch
Purdue University

Bonnie Lawlor
Institute for Scientific Information

John L. Massingill
Dow Chemical Company

Robert McGorrin
Kraft General Foods

Julius J. Menn
Plant Sciences Institute,
  U.S. Department of Agriculture

Marshall Phillips
Office of Agricultural Biotechnology,
  U.S. Department of Agriculture

Daniel M. Quinn
University of Iowa

A. Truman Schwartz
Macalaster College

Stephen A. Szabo
Conoco Inc.

Robert A. Weiss
University of Connecticut

1991 ACS Committee on Corporation Associates

Chairman
Charles S. Sodano
Nabisco Biscuit Company

Morris B. Berenbaum
Allied-Signal Inc.

P.M. Chakrabarti
PPG Industries, Inc.

John W. Collette
Du Pont

Paul D. Klimstra
Searle & Company R&D

Duane S. Lehman
The Dow Chemical Company

Ted J. Logan
Procter & Gamble Company

Kenneth O. MacFadden
W.R. Grace & Company

Patrick J. Marek
Nalco Chemical Company

C. Gordon McCarty
Mobay Corporation

John McShefferty
Gillette Research Institute

Barbara Peterson
3M

Thomas W. Smith
Xerox Corporation

David M. Sturmer
Eastman Kodak Company

David B. Swanson
Engelhard Corporation

Charles S. Tuesday
General Motors Corporation

John C. Wollensak
Ethyl Corporation

ACS Board of Directors Liaison to CAC
S. Allen Heininger
ACS President

ACS Staff Liaison to CAC
Patrick P. McCurdy
Director of Industry Relations

Office of Corporation Associates Staff
Nancy M. Flinn
Special Assistant to the Director of Industry Relations

Brian K. Theil
CA Program Assistant

1991 Mission Statement

The ACS Committee on Corporation Associates represents CA member companies. Our mission is to advise and influence the ACS Board of Directors in decision and policy-making arenas; in the development of programs and services which support technical employees in the chemical and related industries; and in improving the image of chemistry. In addition, Corporation Associates Representatives assure that ACS programs and activities are effectively communicated to CA member companies.

Contents

Contributors   xiii

Preface   xv

EXPLORING CONSULTANT–CLIENT RELATIONSHIPS

1. **Consulting to the Chemical Industry**   3
   Charles H. Kline

   Types of Consultants   4
   Why Use Consultants?   7
   Why Avoid Consultants?   8
   Proprietary Projects   9
   Proposals   10
   Getting Maximum Value   11
   Multiclient Surveys   13
   Syndicated Services   14
   Consulting Personnel   14
   A Look Ahead   16

2. **Industrial Expectations for Consultants and Consulting Services**   21
   Thomas Dykstra

   Criteria for Consultant Selection   22
   Short-Term Consultantships   23
   Long-Term Consultantships   24
   Use of Consultants for Basic and Long-Term Research   24
   Suggested Practices Before and During the Consultant's Visit   25
   Determining Value Added by a Consultant   25
   Potential Barriers to Effective Use of Consultants   26
   Concluding a Consulting Relationship   27
   Conclusion   28

3. **Understanding, Selecting, Managing, and Compensating Consultants   29**
   Howard L. Shenson
   Why Are Consultants Used?   29
   Making the Decision   32
   The Right Consultant for the Job   33
   Consultant Fees   34
   Finding a Qualified Consultant   36
   Evaluating the Potential Consultant   37
   Client Fears   38
   Making the Client–Consultant Relationship Function Smoothly   39

4. **An Academic Perspective on Consulting   41**
   Earl S. Huyser
   Clients and Fields of Chemistry   42
   Consulting Activities   43
   Rewards of Consulting   44

5. **University–Industrial Relationships   47**
   Charles S. Sodano
   STERMPS: What Is It?   49
   How To Start a STERMPS Program   49

6. **Accessing Federal Laboratories Know-How   53**
   Lee W. Rivers
   Role of Federal Laboratories   55
   Enabling Legislation   56
   Barriers to Accessibility   57
   The FLC—Aid to Accessibility   57

CONSULTING OPPORTUNITIES

7. **What Consulting Practices Look Like: The Nuts and Bolts of Organizing a Practice   65**
   Michael Curry

8. **Defining and Marketing Your Consulting Specialty   83**
   Peter R. Lantos
   Advantages and Disadvantages   84
   Finding Clients   85

9. **Opportunities for Retired Chemists   95**
   David G. Bush and John R. Thirtle
   What's Going On?   96
   How Much Are You Worth?   96
   Temporary and Part-Time Employment   97
   Some Typical Consulting Firms   98
   Agencies for Part-Time Professional Employment   99
   What Corporations Are Doing   101
   The Trends   101
   Concerns of Employers, Labor Officials, and Politicians   102
   Volunteering   102
   Volunteering with Financial Support   103
   Some Volunteering Opportunities Requiring Technical Input   104
   Some Volunteer Teaching Opportunities   104
   Some Federal and State Regulations   105
   Summary of Consulting Income   107

10. **Using Consultants To Interpret Regulatory Initiatives   111**
    Mike McCormack
    Organizational Steps   111
    Market Characteristics   113
    Marketing Methods   114
    Insurance Problems   116
    Examples of Work   116
    Organizational and Operational Essentials   118

11. **Major Chemical Company Retirees as Consultants and Market Developers   119**
    Robert W. Belfit, Jr.
    Formation of the Company   120
    Work Performed   122
    Other Ventures   125
    Complementary Products   126
    Quality in Performance   127

12. Robotic Servicing on the Space Station Freedom   131
Dale S. Schrumpf

Purpose   131
Productivity in S.S. Freedom   132
Purpose of Robotic Servicing   132
Types of Robotic Systems   134
Marketing Your Services   136
Hiring a Consultant   137
Conclusions   137

13. Consultation in Sensory Evaluation   139
Gail Vance Civille

14. Chemical Information Consultants: Selection and Vitalization   143
Robert E. Maizell

Professional Qualifications for Information Consultants   145
Services Offered   145
Care and Feeding of an Information Consultant   146
When To Consider Hiring an Information Consultant   147
Selecting and Hiring an Information Consultant   149
How To Control Consulting Costs   151
Potential Sources of Information Consultants   152

INDEX

Index   **157**

Contributors

Robert W. Belfit, Jr.
Omnitech
2715 Ashman Street
Suite 100
Midland, MI 48640

David G. Bush
David G. Bush, Associates
147 Landing Park
Rochester, NY 14625

Gail Vance Civille
Sensory Spectrum, Inc.
24 Washington Street
Chatham, NJ 07928

Michael Curry
941 St. Marks Avenue
Westfield, NJ 07090

Thomas Dykstra
Eastman Kodak Company
Building 83
Rochester, NY 14650-02212

Earl S. Huyser
Department of Chemistry
5029 Malott
University of Kansas
Lawrence, KS 66045

Charles H. Kline
PanGraphion Inc.
389 Ski Trail
Kinnelon, NJ 07405-2247

Peter R. Lantos
The Target Group
1000 Harston Lane
Philadelphia, PA 19118

Robert E. Maizell
Technology Information
  Consultants
5 Science Park
New Haven, CT 06511

Mike McCormack
Institute for Science and Society
Central Washington University
Ellensburg, WA 98926

Lee W. Rivers
Technical Transfer Department
University of South Carolina
Spartanburg, SC 29303

Dale S. Schrumpf
Lockheed Missiles & Space
  Company, Inc.
Building 580
1111 Lockheed Way
Sunnyvale, CA 94089

Howard L. Shenson
20750 Ventura Boulevard
Suite 206
Woodland Hills, CA 91364

Charles S. Sodano
Analytical Methods
Nabisco Biscuit Company
R. M. Schaeberle Technology
  Center
100 DeForest Avenue
East Hanover, NJ 07936-1943

John R. Thirtle
105 Conifer Lane
Rochester, NY 14622-1003

Preface

Trends in Chemical Consulting: Present Opportunities and Future Direction was the title of an ACS Corporation Associates Annual Symposium that was held in conjunction with the ACS National Meeting in Los Angeles in September 1988. The chapters in this book, for the most part, derive from the symposium papers but have been updated and expanded.

The three-part symposium program covered industry's expectations of consultants, how to start a consulting business, and examples of current and future consulting fields. Charles H. Kline, founder of Kline & Company, Inc., international consultants, and president of Pangraphion, was the keynote speaker. He addressed the nature of consulting in the chemical process industry.

This book is based on that symposium; the authors have prepared chapters that provide valuable insight into chemical consulting as a career. Each author writes from personal experience about his or her own technical expertise. The authors range from founders of large consulting firms to individuals who consult as solo practitioners. Some of the chapters provide general overviews of what the client and consultant should expect and receive from a consulting arrangement. Others are specific examples of newer areas that represent opportunities for chemical consultants.

We offer this book to those of you who may be considering consulting as a second career, as a dual career, or as a way to stay active in your field after retirement. Is consulting for you? This book won't provide an absolute answer to that question, but it will give you the benefit of a lot of hard-earned experience.

Acknowledgments

The parent symposium was sponsored by Corporation Associates (CA). Founded in 1952, CA is the formal link between the chemical process industries and the American Chemical Society. More than 100 leaders in the chemical process industries actively support the science and profession of chemistry through participation in this program. Their mission is to advise and influence the ACS Board of Directors in decision and policy-making arenas; in the development of programs and services that support technical employees in industry; and in improving the image of chemistry.

We give special thanks to Nancy Flinn of the American Chemical Society for assistance in the preparation of the symposium and of the book.

We also thank Benjamin J. Luberoff, editor of *CHEMTECH*, for his participation and support of the symposium and for referring a manuscript originally submitted to his magazine for inclusion in this book.

CHARLES S. SODANO
Nabisco Biscuit Company
R. M. Schaeberle Technology Center
East Hanover, NJ 07936–1943

DAVID M. STURMER
Eastman Kodak Company
Corporate Research Laboratories
Rochester, NY 14650

EXPLORING CONSULTANT–CLIENT RELATIONSHIPS

Chapter 1

Consulting to the Chemical Industry

Charles H. Kline

The U.S. chemical industry spends perhaps $1.5 billion a year on consulting services of all types. Many of these services are common to all industry, such as economic forecasting, employee benefits, financial services, executive search, and organizational consulting. These services do not require specialized chemical knowledge.

But other types of consulting do require broad knowledge of chemistry and chemical engineering, the languages that one must speak to understand a chemical company and its problems. Consulting services that require this knowledge generally involve marketing, strategy, and acquisition and divestiture planning on the business side, and technology assessment, research and development, process engineering, environmental protection, and product and environmental safety on the technical side. Work of these types I designate as chemical consulting as opposed to more general management and business consulting. The volume of chemical consulting probably approaches $500 million a year in the United States.

An earlier version of this paper appeared in CHEMTECH, Volume 20, Number 6, June 1990, pages 348–352. Copyright © 1990 by the American Chemical Society. All rights reserved.

2106–5/91/0003$06.00/0 © 1990 American Chemical Society

Types of Consultants

Chemical consultants, the people who provide these services, are of many types.

University Professors. Professors were the first consultants, starting 50 to 100 years ago. They are still prominent today, particularly scientists and engineers as advisers to corporate research and development organizations in such fields as biotechnology and advanced materials. More recently, members of business school faculties have been increasingly used as advisers in strategy and marketing. Most university professors work alone as individual consultants, but they benefit from the cross-fertilization of ideas in the university environment.

Solo Practitioners. Several hundred people in the United States consult as individuals, generally assisted as needed by a full- or part-time secretary and a few temporary employees, often other solo practitioners. Solo practitioners lack the resources of organized firms, but are often expert in narrow fields, for example, production of a specific polymer like polyvinyl chloride. They retain total independence but lack partners with whom to exchange ideas and discuss problems. Furthermore, they have no backup if illness strikes or vacation time arrives. Many individual consultants, particularly those who specialize in technology, belong to the Association of Consulting Chemists & Chemical Engineers.

Consulting Firms. By far the greatest amount of chemical consulting is done by organized firms, ranging in size from a handful of employees to several thousand. A number of firms specialize in one or more aspects of the chemical industry. Thirty-four of the better known, ranging in size from several hundred employees to fewer than 10, are listed in Appendix A; their activities are summarized in Table I. Most of these firms conduct marketing and strategic studies, either across broad sections of the chemical industry or in such specific sectors as petrochemicals, plastics, or process technology. Identification and appraisal of target companies for acquisition is an important activity in many firms.

The two largest firms in this group, each with more than 100 full-time employees and 80% or more of its work devoted to

Table I. Principal Fields of Specialized Chemical Consulting Firms

Field	Number of Firms
Marketing, strategy, and technology in many products	14
Petrochemicals	5
Plastics and packaging	5
Coatings and adhesives	2
Economic data and forecasts	2
Agricultural marketing research	2
Other	4
Total	34

chemicals, are Chem Systems Group, Inc., and Kline & Company, Inc. Chem Systems is an international firm active in strategic planning, diversification studies, mergers and acquisitions, and technical and economic analyses. Its topics range from petrochemicals and processes to polymers and specialty chemicals. Kline & Company, Inc., also an international firm, conducts strategic, marketing, and technical studies and acquisition appraisals in chemicals, particularly specialty and fine chemicals, tonnage inorganics, plastics, advanced materials, and consumer chemicals. A subsidiary, Findtech Inc., specializes in divestiture and acquisition deals.

Another large firm is Springborn Testing Institute Inc., a plastics testing and development laboratory with some consulting activity.

Recently a number of companies have been formed to take advantage of the skills of personnel retired from major chemical companies. Such firms have a few permanent professionals plus a number of part-time consultants on call for specific assignments.

Two of the many large general management consulting firms, Booz Allen & Hamilton, Inc., and McKinsey & Company, Inc., have well-recognized chemical practices, particularly in organizational studies. Of the firms with both management and technical consulting staffs, Arthur D. Little is the leader in the chemical industry. SRI International also offers both technical and management consulting and produces extensive data bases. Battelle Memorial Institute is strong in technical consulting but not in management issues. Table

II summarizes activities of these companies. Although these firms are large in total, their staffs for chemical consulting are smaller than those of some specialized firms.

Table II. Large General Consulting Firms with Chemical Practices

Firm	Management Consulting	Market Data	Technical R&D
Battelle Memorial Institute	—	—	XXX
Booz Allen & Hamilton	XXX	X	—
Arthur D. Little	XXX	X	XXX
McKinsey & Company	XXX	X	—
SRI International	XX	XXX	XXX

NOTE: XXX means high activity, XX means moderate activity, and X means some activity.

The Big Six accounting firms, formerly the Big Eight, have become formidable competitors in general management consulting but thus far have not developed strong chemical practices. Although skilled in nontechnical matters, they appear mystified by the intricacies of chemistry and the life sciences. This group now includes Arthur Anderson & Company, Deloitte & Touche Company, Ernst & Young, KPMG Peat Marrick, and Price Waterhouse.

A third group of firms focus on testing, product safety, health, and environmental protection. The various regional research laboratories fall into this group. For example, among their other activities, Midwest Research Institute is active in screening anticancer drugs and hazardous waste disposal, Southern Research Institute in environmental sampling, and Southwest Research Institute in problems related to the gas industry.

A fourth group of companies provides consumer marketing research on products of all types. A.C. Nielsen Company, Market Facts, Inc., Opinion Research Corporation, and others cover all products. Doane Marketing Research, Inc., and Maritz Marketing Research, Inc., have strong positions in agricultural marketing research.

Finally, some manufacturing companies offer consulting services. For example, E. I. du Pont de Nemours and Company, Inc., has a broad array of services in plant safety, all developed from its own experience. The development of expert systems, sometimes called artificial intelligence, in the application of chemical products may widen considerably the role of manufacturers' consulting services.

Why Use Consultants?
Chemical companies use consultants for a variety of reasons.

Expertise. Good consultants have specialized skills that a company's employees may lack. For example, as a result of their work for many companies, the large general management consultants have become expert in corporate organization and strategy. Similarly, many smaller consultants are skilled in strategies for particular markets or technologies—for example, Chem Systems or Wright Killen & Company, Inc., in process technology; Dewitt & Company, Inc., in petrochemicals; Kline & Company, R.M. Kossoff & Associates, Inc., and Philip Townsend Associates, Inc., in plastics; Kline & Company, in specialty and fine chemicals; Strategic Analysis Inc. (SAI) in catalysts; or the Commodities Research Corporation in fertilizers and other commodities.

Independence. As outsiders, consultants offer independent views. They have nothing personal to gain from their recommendations; they will not be promoted within the client company, otherwise rewarded, or discharged. The best they can hope for is to do such a good job that they will be invited back to conduct another project. Consultants' independence is particularly useful to chief executives who need an impartial foil with whom to discuss questions of internal organization and personnel. It is also helpful in developing material for government regulatory agencies or expert testimony for lawsuits.

Impartiality. Because of their independence, consultants often serve as arbiters to settle disputes between warring factions in a company, for example, research and marketing. Neither faction will trust the other, but each will accept the recommendations of an impartial outsider.

Focus. Resolving business problems takes time, a commodity often lacking among executives. Consulting attracts analytically minded problem-solvers. Free of operating responsibilities, they can analyze the client's problem in great depth and carry it through implementation of recommendations.

Creativity. From their experience with many clients, consultants develop advanced concepts that they can apply to the specific problem at hand. Examples of such concepts are McKinsey & Company's work on the Technological S-Curve and Value in Use and Kline & Company's development of the Differentiation Index and its Classification of Commodity, Fine, and Specialty Chemicals.

Anonymity. When a chemical company needs to pursue the early stages of a project in secrecy, consultants can provide the necessary anonymity. They can investigate the potential market for a new product, determine a company's image versus those of its competitors, or analyze the operations of a target for acquisition without disclosing the identity of the interested party.

Rapid Response. Consultants can often respond quickly to requests for data or other information not available through a client's internal sources or computerized data banks.

Reduced Costs. In the lean and mean days of the early 1980s, chemical companies found they could make substantial savings by reducing permanent internal staffs for planning, marketing research, and project analysis and instead using consultants as needed on a temporary basis. Similarly, chemical companies find it more economical to subscribe to consultants' multiclient surveys and syndicated services than to develop such large bodies of data themselves.

Cachet. Some companies use prestigious consulting firms to give their cachet of approval to unpopular or radically new actions—for example, a corporate restructuring, sale of a division, or the dismissal of a top executive.

Why Avoid Consultants?

Consultants also have disadvantages. First, although effective in reducing costs and increasing sales and profits, at the beginning they do represent an out-of-pocket cost. Second, busy executives must take time to familiarize the consultants with the company and its problems, particularly on a first assignment. Third, introduction of consultants to undertake organizational, personnel, or other changes sometimes causes resentment among the firm's personnel.

Proprietary Projects

Most consulting assignments are custom-designed to address a specific topic, either raised by the client's executives or proposed by the consultant. Often the topic is a problem that has worried the client company for several years but which the company expects the consultant to solve in three months.

From the consultant's point of view, the best clients are sophisticated companies with strong internal capabilities in staff work. In such firms the key executives understand the value of information and analysis, but they do not buy services they do not need. The clients also know how to use consultants and what to expect from them. The clients have their own short list of consultants competent in the area under investigation—for example, polymer additives or biocides—and the necessary skills in research and development, distribution, acquisition, or other functions.

The consultant and the client alike must fully understand the business or technical problem. The client can help by providing all relevant background information. The consultant can help by strategic questioning. Even sophisticated clients often do not know what they really need. Perhaps the consultant's greatest contribution is to identify the real nature of the problem and the steps that can be taken to solve it—a contribution for which the consultant is rarely paid.

The client company should describe the business decision it proposes to make as a result of the consulting project, when the decision must be made, and how much the action to be taken will cost. For example, the decision to build a new grassroots plant costing $100 million may warrant a very detailed 6-month study of the market, process economics, competition, and timing in the economic cycle. On the other hand, the decision to launch an 18-month research project may warrant only a 2-week appraisal of these factors.

The consultant, in turn, must give some indication of the cost of the project. At first, only an estimate is required. A definite quotation can be given later when the proposal is submitted. Billings for individual consultants vary widely with their age and experience, but generally fall in the range of $500 to $2500 a day plus out-of-pocket expenses.

Successful relationships between client and consultant are founded on mutual trust. The client company must take the consultant fully into its confidence. No reputable consultant will betray the client's secrets. The company should reveal what it already knows so that the consultant does not seek to reinvent the wheel. Occasionally a consultant is retained merely to check the accuracy of information already in hand. Although sometimes necessary, such projects are boring to consultants and may irritate the client's employees.

Proposals

After preliminary discussions, often spread over several sessions, the consultant prepares a proposal to conduct a project for the client, generally including these items:

- the nature of the problem
- the objective of the proposed project
- the work to be done
- the dates of completion and interim reviews
- the reports, oral summaries, or other presentations of the results to be given
- the need to maintain anonymity of the client in outside contacts
- consulting personnel to be assigned to the project, their backgrounds, and experience
- subcontract services to be used, if any
- charges, generally split between fees for services and travel, subcontracts, and other out-of-pocket expenses
- client safeguards such as confidentiality and no conflict of interest

- the consultant's limitation for damages
- a space for acceptance of the proposal

Proposals vary from simple three- to four-page letters of agreement to elaborate documents running up to 100 pages or more. The longer ones are expensive and time-consuming to prepare. The client company should have sufficient advance knowledge of the various consultants to limit its requests for proposals to two or three firms. A greater number wastes the client's time in evaluating the proposals and, of course, the consultants' time in preparing them.

In requesting and accepting proposals, clients should match the consultant to the project. Some projects require the depth of talent and breadth of experience of a large firm with its overseas offices, foreign language skills, communications facilities, and data banks. Other projects need only the experience and good sense of one consultant. One of the largest fiber producers in the United States has most of its consumer research processed by two homemakers in its headquarters town.

An important item in the proposal, of course, is the price of the project. For projects that can be accurately defined in advance, a specific price can be quoted. For less clear-cut projects, such as organizational studies, the price will often be quoted on a per diem basis, the total not to exceed a fixed maximum.

Particularly important in accepting or rejecting a proposal are questions of personnel. Who will actually do the work on the project? Will they be on the consultant's permanent staff, temporary employees (stringers), or employees of a subcontract service firm? What are their backgrounds? Who will supervise the project? Will the supervisor actually be involved in the work or merely appear at the initial sales meeting and the final presentation?

Getting Maximum Value

Once a client company has accepted a consultant's proposal, it naturally wishes to get the maximum value from its often considerable expenditure. To do so, the company should:

1. Assign one senior staff member, with the authority to make

decisions, to work with the consultant through the entire assignment. Don't change horses in midstream.

2. Whenever possible, allow reasonable lead time before work begins. The best person to conduct the project may be just completing another job. Insistence on starting at once means that the best personnel available at the time—not necessarily the best people for the job—will be assigned to it.

3. Monitor the work of consultants, but do not overburden them with meetings and telephone calls to the point that little or nothing gets accomplished.

4. Encourage the consultant to develop original ways of analyzing the business problem at hand and to produce unique concepts and analyses that the client company would not have conceived itself.

5. Prevent project drift, the tendency to let the project slide away from the original objective to other goals. In an informal poll of consultants this was the difficulty most frequently cited in client relations.

6. Avoid written reports wherever possible. If the results will be used internally, have them presented orally with slides or transparencies summarizing all findings. Copies can be provided for later study. Of course, written reports are necessary when the findings must be circulated to outsiders.

7. Don't shoot the bearer of bad tidings. Many (if not most) consulting projects give negative results: Don't build the plant; don't change the organization; don't make the acquisition. Such findings are equally as valuable as positive results, even if less interesting. For this reason the executive who commissioned the project may not be the best judge of its worth; he may himself be too deeply involved to be objective.

Multiclient Surveys

Many consultants issue technical or market studies of particular topics for sale to a number of different clients. These multiclient surveys, first offered by the late Roger Williams, give a broad overview of a market or technology, current trends, present and prospective suppliers, and the forecast position five or more years in the future. They analyze these factors from a single point of view at a single point in time. Accordingly, they serve as the basis for differentiating the product, segmenting the market, or developing other strategies. Findex (Bethesda, Maryland) issues a directory of multiclient surveys on U.S. markets, and Marketsearch (London) issues a similar listing of international surveys.

Multiclient surveys typically cost the consulting firm $100,000–$750,000. However, as this cost is spread over 10–50 subscribers, the price to each company is generally $10,000–$25,000. (Various specialty publishers offer far less comprehensive analyses at $500–$2500 each.) The total cost to the consultant is typically 10–25 times the subscriber's price. These surveys thus offer a considerable value to companies already active in a given market or technology and particularly to companies considering entry into the field.

If a company is interested in the topic of a multiclient survey, it pays to subscribe early, before the detailed work has begun. At this point, with only a handful of subscribers, the consulting firm can modify the project to take each client's interest into account. Later the subscriber must take what's on the shelf. It pays also to take full advantage of the consultant's offering: oral presentations of the findings to the subscriber's management, access to the raw data, and (where available) computer diskettes of the results that can be updated in the future.

Because multiclient surveys cover broad topics, they may not give sufficient detail on individual niches. Frequently a subscribing company follows up the multiclient study with a more specific proprietary project—for example, to develop a marketing strategy for a new product.

Syndicated Services

Some consultants issue periodic data or reports on the broad economy or particular products or markets. Like multiclient surveys, these syndicated services spread the cost of producing a mass of data over a number of subscribers. They supply much information that companies formerly compiled within their own organizations but can no longer produce with today's reduced staffs. Subscription prices run $7,000–$25,000 a year.

Consulting Personnel

A good consultant is a problem-solver with an analytical mind, the patience to probe deeply into a problem, and the ability to communicate the findings convincingly both orally and in writing. Although having special expertise in a few fields, the consultant also must have the confidence and flexibility to take on a wide range of assignments.

Consulting is hard work. It involves extensive travel, constant deadlines, and difficult mental shifts from project to project. Results are often negative, indicating that the client should not undertake the proposed action—sound recommendations, but discouraging. The consultant sometimes faces the hostility of a client's employees, who feel threatened by the work or disagree with his or her results. And even after a project is complete, the consultant may not learn until long afterward whether the recommendations were accepted and how successful they proved in practice.

Why then do people become consultants? The principal reason is the attraction of an intellectually stimulating professional career. Most consultants would prefer to be leading experts in their fields supervising a few other professionals, rather than corporate executives managing thousands of employees. They like the peripatetic, problem-solving life. And they take pride in generalizing from their experience on many individual projects to develop new principles of management.

A second reason is the breadth of exposure to industrial problems; in fact, some junior chemists or engineers join consulting firms as a step to higher positions in industry. A third reason is independence. Consultants must work on the projects their firms assign, but they have considerably more freedom to choose these projects and to develop their own ideas than in the typical client firm.

Many people believe that consulting is a game for the young. The large general management consultants almost exclusively hire people under 30 and retire them at 50 to 60. They prefer MBA graduates from the major business schools, preferably with some work experience, and (for their chemical practices) an undergraduate degree in science or engineering. Successful consultants of this background usually become partners of their firms in their early- to mid-30s.

The more-technical management consultants, such as Arthur D. Little and SRI International, are more inclined to hire a few older people with greater experience in the industry. They have a more mixed group—MBAs, M.Sc.s, and Ph.D.s in various disciplines. In the medium to small consulting firms, practice varies. At Kline & Company, Inc., for example, we hire mostly younger professionals with technical degrees, generally with several years' working experience, and usually an MBA. At the other extreme, some firms, such as Condux Inc., Monkman International Consultants, and Omni Tech International Ltd., hire primarily retirees and other experienced mature people.

Solo consultants generally enter the business at 40 to 50. By this age they have acquired enough practical experience and established enough contacts in the industry to justify setting up a practice. Some solo consultants expand into sizable firms by putting their earnings back into the business to hire additional people and to buy the increasingly sophisticated equipment required. Others choose to remain one-person shops.

Recently many corporate executives considering (or forced to take) early retirement have looked to consulting as a second career. As Howard Upton wrote in *The Wall Street Journal,*

> Breathes there a manager with soul so dead
> Who never to himself hath said,
> "Corporate life's become so repugnant
> I think I'll quit and become a consultant"?

In November 1988, the American Chemical Society sent its senior members an optimistic article on consulting as a new career for retirees. In my observation, retirees older than 50 find it difficult to build a successful consulting practice. Their experience has often made them expert in narrow niches of interest to only a handful of potential clients. Frequently, they lack the ability to promote themselves and sell their services. And, frankly, many retirees will not make the strenuous effort necessary to establish themselves in a demanding profession.

A Look Ahead

Consulting to the chemical industry, both on general management and, more specifically, on chemical topics, is increasing at 20–25% a year. The restructured chemical industry is more willing to seek outside help than in the past. Many people in the industry have learned that the herd instinct makes for poor strategy. Rather than follow the industry trend, the large traditional chemical companies are using consultants to search out unique approaches to future growth. Similarly, smaller emerging firms are recognizing the value of specialists experienced in the industry.

The business is becoming more and more international. Overseas revenues of major firms are growing at twice the rate of domestic income. It is essential to be in Europe and Japan; other locations, though less critical, are helpful. Regional management from multiple centers is ending. Consulting firms increasingly run worldwide practices in their various specialties, each directed from a single center.

From a low-capital cottage industry consulting has evolved to a much more capital-intensive business. When I started Kline & Company, Inc., in 1959, I had a typewriter and an adding machine (both mechanical), a slide rule, and the CRC Press's *Handbook of Chemistry and Physics*. By 1988 our U.S. office alone had 50 personal computers, a minicomputer with 25 terminals, six printers, computer graphics equipment, two facsimile (fax) transmitter–receivers, a telex, and five copiers. And our library received more than 300 periodicals, had access to numerous computerized data bases, and contained several thousand bound volumes and large quantities of unbound data.

This trend to higher capitalization will continue as new data bases and new equipment for accessing and processing information become available.

On the client side, more and more manufacturers are making a senior manager a joint participant in consulting projects. This step raises the quality of the project, increases the information left behind after its completion, and helps implement the recommendations. Such close collaboration eliminates the need for the ponderous written reports that consultants formerly produced. Project results today are typically presented in oral sessions, with slides or transparencies. The report is merely a short executive summary plus copies of the slides and relevant backup data.

The principal focus of consulting through the 1990s will be positioning chemical companies to maintain superior competitive stances worldwide. If consultants do their work well, they will help produce a stronger, more effective, and more profitable chemical industry throughout the world. Let us hope that both consultants and manufacturers succeed in this endeavor.

Appendix A. Specialized Chemical Business Consulting and Service Firms

Company	Location	Product Areas	Size Group[a]	Chemical Work (%)
Catalytica Associates[b]	Mountain View, CA	Primarily research	C	100
Charles River Associates[c]	Boston, MA	Various	B	35
Chem Systems Group[c]	Tarrytown, NY	All industrial	B	85
Chemical Data	Houston, TX	Petrochemicals, fuels	F	100
Chemical Market Associates[c]	Houston, TX	Petrochemicals	F	100
Chemical Marketing Services[c]	Cincinnati, OH	Coatings, adhesives	E	100
Commodities Research Corporation[c]	New York, NY	Fertilizers, other commodities	B	35
Condux Inc.	Newark, DE	All industrial	C[d]	90
Consulting Resources Corporation	Lexington, MA	All industrial	E	100
Data Resources Inc.	New York, NY	Economic forecasts	A	10
DeWitt & Company[c]	Houston, TX	Petrochemicals	D	100
Doane Marketing Research	St. Louis, MO	Crop and animal chemicals	C	50
Einhorn Associates	Milwaukee, WI	Coatings	F	70
Gorham Advanced Materials Institute[b]	Gorham, ME	Advanced materials	E	100
Gorham International	Gorham, ME	Paper chemicals	F	15
Harborside Research Associates	Salem, MA	Polymers, packaging	F	100
Colin A. Houston & Associates	Mamaroneck, NY	Surfactants, other	E	100
Hull & Company	Greenwich, CT	All industrial	E	100
Innotech Corporation	Trumbull, CT	Creative innovation	D	65

Kline & Company, Inc.[c]	Fairfield, NJ	All industrial; also consumer chemicals	B	85
R.M. Kossoff & Associates	New York, NY	Plastics	E	100
Maritz Marketing Research Inc.	St. Louis, MO	Crop and animal chemicals, equipment	A	10
Monkman International Consultants[c]	Wilmington, DE	All industrial	F	100
Omni Tech International Ltd.	Midland, MI	All industrial	B[d]	70
Pace Consultants	Houston, TX	Hydrocarbons	D	25
Probe Economics	Mt. Kisco, NY	Forecasting	E	95
Purvin & Goetz	Dallas, TX	Petrochemicals, oil and gas	B	10
Science & Technology Corporation	Bartlesville, OK	Chemicals and engineering	B	20
Strategic Analysis Inc.[c]	Reading, PA	All industrial	C	70
Skeist Inc.	Whippany, NJ	Polymers	E	100
Springborn Testing Inst.[b]	Enfield, CT	Plastics (mostly testing)	B	100
Technomic Consultants Inc.	Deerfield, IL	All industrial	D	35
Philip Townsend Associates	Houston, TX	Plastics	D	90
Wright Killen & Company	Houston, TX	Process industries	E	50

[a]Total employees: A, more than 500; B, 101–500; C, 51–100; D, 26–50; E, 11–25; F, 10 and fewer.
[b]Maintains laboratory.
[c]Has offices overseas.
[d]Primarily part-time retirees.

Chapter 2

Industrial Expectations for Consultants and Consulting Services

Thomas Dykstra

The observations in this chapter are restricted to technical (as opposed to manufacturing or business) consultants coming primarily from universities. They are not full-time consultants, but university professors for whom consulting is a sideline. Information for this chapter was generated by sending questionnaires on consulting relationships to 50 scientists and managers in Kodak's research laboratories who have substantial interest in such relationships. The questionnaire was followed by focus group discussions with the respondents.

Scientific disciplines of consultants used by the respondents included those given in the list on page 22. This diverse list represents only part of the total number of fields for which consultants are brought to Kodak's research laboratories.

It became obvious during the study that working scientists have a strong sense of ownership of the technical consulting relationship. Though management has a role in approving, funding, and administering consultant contracts, the communication, credibility, confidence, and trust between scientist and consultant must be nurtured

Some Scientific Disciplines of Consultants

bacteriology	molecular biology
biochemical engineering	molecular design
biochemistry	mycology
chemical engineering	optics
chemical physics	organic chemistry
chemistry	pharmacology
clean-room procedures	photoelectronic properties
cloning	of solids
compound semiconductor	physics
devices and materials	physiology
digital image processing	polymer science
inorganic chemistry	solid-state chemistry
marketing	solid-state physics
materials science	surface and colloid science
mechanics	synchrotron radiation sourcing
microbiology	theoretical chemistry
microelectronics	vacuum coating
	vibrational spectroscopy

and protected. In addition, scientists should have the primary role in choosing consultants.

Criteria for Consultant Selection

Weighted criteria of consultant selection appear in Table I. Obviously, knowledge of the field is paramount, but interpersonal skills are also important. In discussion, we noted that the consultant should have a good scientific network in the field. Although an impressive scientific reputation, a prodigious publishing record, and prior acquaintanceships are helpful, they are not sufficient in themselves. Big egos should be avoided. It is inadvisable to develop a list of consultants based on prestige and aimed at somehow improving the organization's reputation. Bright people relatively new to academe and hungry to develop their professional status should be strongly considered. A source of such people is the Presidential Young Investigators, who have been selected by the National Science Foundation for funding to match that from industry.

Visits by, or to, a consultant candidate, and even collaboration on a short project, are good activities preliminary to signing a

Table I. Weighting of Selection Criteria

Criteria	Average Weighting by Respondents[a]
Knowledge of field	9.4
Aptitude for stimulating creative discussion	7.8
Knowledge of specific problem	7.1
Possibility of collaborative research	7.0
Ability to provide background and reference data	6.8
Source of hiring candidates	4.3
Source of facilities and equipment	3.8

[a] 10 = high; 1 = low.

consulting contract. All parties should be clear about availability and potential conflicts of interest. Information about potential consultants should be verified.

Short-Term Consultantships

Advantages. Short-term (one to two days) arrangements are a good introduction to a longer term contract. They carry little commitment by either party. The consultant, lacking long association with the company, is unencumbered with company culture. Short-term agreements allow specific expertise to be applied to specific problems. Alternatively, from several short-term visits, one can obtain a variety of ideas and perspectives on a problem. If an area of technology is new to a company, some short-term consultants can bring a company up to date quickly in the area.

Finally, if several short-term consultants have good reactions to the company and its research, their reactions are spread to several university departments.

Disadvantages. When the relationship is short term and each party then goes its own way, confidentiality may be a worry on all sides. It should be adequately protected in writing, and an element of trust, even in the short term, should be present.

Lack of background in the company may leave the consultant unencumbered by the culture, but it can be detrimental to the depth and relevance of the consulting. For example, much time may be spent in thinking up and proposing solutions that have already been tried by the company. Also, factors such as company culture, financial conditions, and policy often affect the path a company takes to solve a problem.

Finally, lack of adequate time and continued contact may limit the depth and creativity of the short-term consulting process.

Long-Term Consultantships

Advantages. In long-term consulting, the consultant learns the culture of the company and becomes acquainted with its people, technology, and programs. This knowledge can be very helpful to the giving of advice and to problem solving, as previously noted. Comparison with similar research done outside the company and review of the internal program can also be performed by the consultant.

With time, trust between parties and a level of ownership of the technical problems by the consultant develop.

Finally, other interactions become possible. They include equipment and facilities exchange, joint research, and awareness of and access to good graduate students as hiring candidates.

Disadvantages. As technical interests and problems change, the long-term consultant may lack the specific expertise required.

In some cases, the growing relationship may somehow grow to include the university, a prospect that may not always be welcomed.

Very long-term consultancies may nurture a tendency toward complacency and worn-out ideas. A good-old-boy syndrome may develop with a resulting lack of rigor. Then, too, long-term relationships may be difficult and painful to terminate when it becomes advisable or necessary.

Use of Consultants for Basic and Long-Term Research

Basic Research. In addition to applicable comments in previous sections, consultants are useful for influencing management to invest in new or needed research. Consultants provide an excellent window on related research in academe. In general, consulting relationships for basic research are easier to administer and are less constrained by proprietary issues.

Long-Term Research. Consultants in long-term research are most valuable in the early stages of a program. They help provide the fundamentals of the science or technology and good ideas on which to begin working.

With time, because of the concentrated effort a company can apply to a problem, the client tends to outrun the expertise of the consultant. Then, as complex systems grow out of the program, the consultant finds it difficult to keep pace. Thus, during the life of a program, the company may need to change to using a different type of consultant.

Suggested Practices Before and During the Consultant's Visit

It is best if an agenda and briefs of the problem or project are sent to the consultant well before the visit. This is sometimes difficult to do, particularly in a large company with a matrix organization. At a minimum, a schedule of the next day's discussions should be at the consultant's hotel or office the night before the visit.

Many organizations begin the visit with a group meeting where the consultant can review current activities in the technical area. Preferred practice is to provide the consultant a comfortable conference room, avoiding the need for the consultant to walk extensively around the site to meet the clients. Each discussion should be long enough to provide for in-depth discussion and time to let the dynamics of the creative process occur. Often, the visit concludes with another group meeting to review the discussions and to determine what direction the work should take.

Determining Value Added by a Consultant

More or less measurable ways in which one might determine the value added by a consultant follow:
- performance against preset goals and criteria
- analysis of value derived from consultant versus that of other investments
- resulting new projects
- changed procedures

- unsolicited testimonials
- fulfillment of the schedule
- avoidance of telling us only we want to hear

Subjective or more visceral methods are described by comments, such as "You know value when you see it" and "If one has to ask, value is doubtful."

Potential Barriers to Effective Use of Consultants

Confidentiality. Confidentiality can be overblown as an issue in consulting and at times can be an excuse for not putting a technical program under the scrutiny of an outside expert. It may be a true concern when the relationship is short-term and when the field is competitive. However, if discussions are limited to science and technology and avoid business issues, confidentiality is rarely a barrier.

For the consultant relationship to work well, all parties have to give as well as get information. A beginning professor with a small group may fear that a large industrial effort will innundate his own efforts in an area. On the other hand, in publication of their results, academicians often are not delayed by matters such as patent clearance that may concern their industrial counterparts.

Conflict of Interest. More and more academicians are starting their own businesses based on their research findings and thus could be viewed as commercial competitors. Finally, some fields are so hot (e.g., superconductivity and nonlinear optics) that people working in them are generally reluctant to share recent findings.

Collaboration and joint research are good ways to build trust and communication when the industrial client and the consultant's research areas are closely aligned.

Too Much or Too Little Informal Interaction. A balance should be struck between developing open and trusting rapport with a consultant while at the same time allowing the consultant to render and maintain an objective viewpoint. If friendships become too close, the consultant may become reluctant to offer objective criticism when it is required.

Insufficient Offsite Involvement by the Consultant. If the consultant considers the client's problems only while visiting the site, the relationship is unlikely to be useful. On the other hand, the professor's time is in short supply and valuable. The client cannot expect unlimited free access, particularly if the consultant is not on an annual retainer. Some consultants have charged hourly for offsite time and time spent in telephone conferences and other activities on behalf of the client. Two solutions are collaborative work and the provision of research grants by the client. Generally, long-term consultantships provide for better continuity between visits.

In any event, the client should always remember to be as specific as possible in between-visit requests.

Other Barriers. Other potential barriers to effective use of consultants are

1. a consultant who is either too quiet and shy or too impulsive and bombastic
2. a contract that is primarily defensive and is continued only as a means of denying access to the consultant by other companies
3. a contract that contains unnecessary exclusivity requirements
4. an atmosphere in which the consultant becomes overly desirous to please

Concluding a Consulting Relationship

Concluding a consulting relationship works well under the following conditions:

1. There is a mutually recognizable program change.
2. Expectations and renewal terms are clear from the beginning.
3. Adequate notice of ending the relationship is given.
4. The consultant has other demands on his or her time.
5. Open and objective evaluation is done regularly by both parties.

6. Renewal is not automatic or done with little review.
7. Tact, sensitivity, and diplomacy are employed.
8. The consultant's prestige and reputation are not threatened.
9. The contract is not a defensive one (see 2 under Other Barriers).
10. Other interactions (with students, facilities, and the university) are not complications.
11. Communication and negotiation with the consultant remains with the scientist and is not usurped by management.

Conclusion

Consultant relationships can take different forms, depending on the needs of the client. Good analysis of need, careful selection, open communication, and the development of trust between parties all contribute to a successful and profitable relationship for the client and the consultant alike.

Acknowledgment

I acknowledge the cooperation and input provided by those who responded to the questionnaire and participated in the focus group discussions.

Chapter 3

Understanding, Selecting, Managing, and Compensating Consultants

Howard L. Shenson

The idea of using a consultant was unusual 20 or 30 years ago. Today it is commonplace, not only in large organizations, but in small ones, where the greatest growth has occurred. Consultants are not the easiest people with whom to work. Often they are described as prima donnas who are difficult to get along with and who march to a different drummer. They are unique individuals, and making the relationship effective and efficient requires as much effort and attention from the client as it does from the consultant. But the effort is a worthy investment. Consultants can, and have, added meaningfully to the success of the projects with which they have become involved.

Why Are Consultants Used?

Consultants are used in specific situations. The best results in the client–consultant relationship come when the client thinks very carefully about the need for and the appropriateness of using the consultant. What are the situations where consultants are used?

Specialized Expertise. The classic, and probably primary, reason for using a consultant is a need for specialized expertise or unique talent. It may be unavailable in the client's organization, or the employee who has the expertise or talent may be occupied with other activities, and the job must be done immediately.

Unbiased Opinions. The second need, and a growing one, is the desire for an independent, unbiased, outside, frank opinion. The client needs someone who can distinguish the trees from the forest, someone who is not so concerned with the political situation within the organization that judgment becomes biased.

Temporary Technical Assistance. The third reason, and the one behind the greatest acceleration in consulting activity in the past few years, is the rapid change and increasing complexity of technology in all fields of endeavor. If a company has a specific technical job to be done, particularly in a limited time, it is often advisable to go outside its organization for the required expertise and talent. Contracting with a consultant is often more efficient and cost-effective than hiring and training an employee, then having the contract completed in short order or canceled and having to terminate the employee, all at considerably more time, expense, and stress than finding the consultant would have cost in the first place. The U.S. government, in its contracting efforts with consultants, particularly through the Department of Defense, has created some very effective models for using technical people in this way, and its methods have been adopted by the private sector of the economy.

Business and Cash-Flow Problems. The fourth reason for retaining consultants is business or cash-flow problems. Organizations tend to retain consultants when they are getting in trouble, and sometimes, of course, they bring consultants in too late. Traditionally, this use of consultants was associated almost exclusively with the private sector. Today, however, business and cash-flow problems are a problem even for nonprofit organizations.

Expertise in Acquisition of Resources. A fifth reason for using consultants is to obtain expert assistance in the acquisition of resources. Classically, we think of headhunters, or executive search consultants, in this regard. Consultants may be useful not only in acquiring personnel, but also in acquiring a variety of other resources, such as raw materials and foreign manufacturing sites or dealers. When a company needs a specific resource of any type, a consultant with the appropriate acquisition skills may be a well-advised investment.

Political and Organizational Problems. The sixth reason that organizations tend to bring in consultants is to handle internal political or organizational problems. Handling such necessary but unpleasant tasks impersonally, leaving management with unsullied hands and a smoothly running organization, is a service that some, but not all, consultants are willing and qualified to perform.

Regulation. The seventh rationale for retaining a consultant is created by the rapidly growing number of government regulations. The more laws that are passed and the more administrative regulations established, the greater is the need for experts in the design of systems and procedures necessary for compliance. In almost all cases, it simply is not possible for internal staff to be aware of all of the ramifications of regulations, and retention of external expertise is often vital.

Sudden Availability of Funds. Although the fact is often not admitted by management, the mere availability of funding typically gives rise to the retention of consultants. A rapid buildup of money seems to make some managers feel an urgent need to put it to use. Often, they will turn to pet projects, ideas they've had on the back burner that can now be activated. Because consultants are available and able to respond quickly, they are viewed as a means of implementation in such situations. This phenomenon is much like walking into a shopping mall with a lot of money in your pocket. The more you have, the more you are likely to spend. Such spending sometimes is the most inappropriate outlay for a consultant.

Handling Key Personnel. The hiring, firing, or relocation of key personnel, often as a result of political motivations, is another task with which some consultants are willing and able to assist. This rationale, obviously, is not unrelated to the political and organizational rationale noted earlier. Because of the increasing liability confronting management in the personnel area, consultants are viewed as being able not only to handle an unpleasant task, but also to insulate management.

Training. Although most large organizations have an internal training capability, they often use consultants to expand the training function. And in smaller organizations, consultants sometimes are the only training resource available. Consultants are in a unique position to provide variety in the training function and to meet a shorter term or one-time need for training in a specialized area. They can free organizational training resources to handle more common and regular needs.

Making the Decision

How do you decide whether it is advisable to use a consultant in your organization? If you have identified one of the 10 situations outlined in the previous section, retaining a consultant is quite often—if not always—the appropriate move. Or you may wish to bring in a consultant for purely economic reasons. A trend in western Europe, in particular, and to a certain extent in the United States, is to recognize the full cost of employees. Such expense involves not only their salaries and fringe benefits, but also such matters as their clerical support, the space they occupy, and the costs of hiring and supervision. Thus, although on the surface a consultant's rates may seem high, the overall savings may be significant.

When you believe you have identified a particular need for a consultant and the possibility of a real economic benefit, be careful. In my experience, and the experience of firms I have worked with and researched, probably one-third of the consulting business in this country would not be undertaken if management took time to review its needs and circumstances and to think logically through its problems. If management does so in a careful and timely manner, a consultant often is not needed. In many cases, management simply delegates responsibility by retaining a consultant just as it would delegate responsibility to someone on its staff. A professional employee of mine once reminded me during an exit interview that the art of delegation entailed more than depositing the entire contents of my desk on his desk. This definitely is a hazard to guard against sometimes when deciding to use a consultant.

The Right Consultant for the Job

How do you select the right consultant for the job? Sound selection requires that the client organization be intimately familiar with its own needs, problems, opportunities, and circumstances. Only then can the client be ensured that it will choose a consultant who meets its precise requirements.

Perhaps the first issue in selection is whether the consultant should be an advisory or operational consultant. Historically, consultants have had an advisory status. They performed whatever research and analysis was appropriate and arrived at conclusions. They made recommendations, usually in the form of a written report, to the client. They then departed and had no further involvement in the outcome. Today, things are different; consultants have become much more operational. Not only do they do the research and analysis, arrive at the conclusions, and make the recommendations, but they also remain to implement the recommendations.

Which method of operation is more appropriate for you? Most clients in most industrial situations today seek operational consultants. They want the consultant to function as project manager as well as expert adviser and to take responsibility for seeing the job through to its conclusion. However, this is not the most effective or efficient use of the capabilities of many consultants. Some consultants function primarily as project managers, whereas others are primarily creative thinkers; those talents do not necessarily reside within the same individual. You must know what you really need and what you are really getting.

A consultant may be a process consultant or a functional consultant. These terms have interesting meanings, and I must be careful when talking to people in the chemical industry about process, but in the terminology of the consulting professional, a process consultant is someone who has a particular skill or expertise. A statistician, for example is a process consultant. The industry involved probably makes little difference to a process consultant's expertise, capabilities, and results. Process consultants are primarily process oriented and work across industry lines. In some situations, process consultants are the right choice because of their intimate knowledge of the technology. In other cases, the right choice is a functional consultant, someone who is knowledgeable within a particular industry. The functional consultant probably uses many different technologies but is very knowledgeable in a particular type of subactivity, for example, within the chemical industry as a whole.

Should the consultant be academically or commercially based? That depends to a certain extent on the client's application. Is the desired result more a function of state-of-the-art knowledge or practical, on-line experience in an industrial setting? In certain situations, an academic is the best solution.

Do you need a large consulting firm, a small firm, or a solo practitioner? You need a firm or individual who is right for the activity in which you are engaged. Many times in working with consultants I have had owners of small firms or solo practitioners say, "Well, IBM would never use me because I'm too small; they use large consulting firms." That is not the case. Granted, IBM is not likely to use a solo practitioner to handle its corporate auditing function. A small firm or solo consultant probably would not have the proper resources for that task. But in highly creative, individualistic areas, a small firm or solo consultant often produces much better results for the client than a large firm that may be more bureaucratic than creative.

So look at the task, and look at the type of consultant who would be most appropriate. There are ways in which even small consulting firms can augment their resources to be most effective.

Consultant Fees

What fees can you expect to pay? I do a yearly national survey of approximately 7000 consultants to find out what fees they are charging, what incomes they are earning, and what operating practices they follow. The results are that on an average, consultants charge more than $1100 a day. That is what they charge a client for a day's service when they are working on a daily billing rate basis. And that is for all kinds of consultants. The rate within the chemical industry is a dollar less than the national average. Work can be done on an hourly rate basis or a daily rate basis. For most consultants, a billable day is eight hours. Consultants determine their fees on the basis of three different charges, which add up to the daily billing rate. The first is the value of their labor, the second is overhead, and the third is their profit—their return on investment, or their return for taking the risk, or their opportunity costs, however you look at profit.

On the surface, $1100 a day seems like a lot of money. Many consultants charge $2000, $3000, or $4000 a day, and some charge only $200 or $300 a day; $1100 is only an average daily billing rate and not everyone is obliged to use it. Although consultants appear to some of their clients to be expensive, a consultant typically does not work a billable day every day, in spite of working hard to have as many as possible. Most consultants work 11 or 12 billable days a month. They spend the rest of the time on professional development, marketing, and other activities associated with running their businesses. They also are entitled to get sick occasionally, have a vacation day, and celebrate holidays.

Many consultants today do not find it in their interest to disclose their fees to their clients on a daily or hourly rate basis. When the consultant answers the client's inevitable question, "What are you going to charge us for this service?" $1100 a day or $140 per hour, or whatever figure is being quoted, seems very expensive. This is particularly so, sometimes, when the executive who is making the decision to retain the consultant erroneously takes the view that the consultant pockets that $1100 or $1200 a day and goes off to live the good life. Then the executive compares the fee with his own salary, and even acknowledging that the consultant is good, wonders just how good the consultant can be.

Running a consulting practice entails costs, like any other business; generally, only 33–40% of what the consultant charges the client winds up in the consultant's pocket as gross pretax income. Because the consultant is billing clients only 40–50% of the time, consultants in general tend to make an average of $90,000–$100,000 per year. Obviously, some do better and some not as well.

Many consultants have become wary of working for an hourly or daily rate. Often they prefer to work for a fixed fee, and this kind of arrangement is almost always in the client's interest. You know what you are going to get and what you are going to pay, and you don't have to worry about the consultant's ability to estimate. All you have to decide is whether the solution to your problem is worth the stated amount of dollars. Usually, consultants also are willing to quote fixed fees plus expenses when it is difficult to estimate direct expenses such as travel.

Another form of consulting compensation is contingency fees. It is not uncommon today for some consultants to work partially or even entirely on a contingency basis. But about 33–40% of the nation's consultants consider contingency fees unethical and will not provide services on such a basis. Those who will work for them will frequently require that some sort of measure of outcome be established, but that is not always possible in a way that will be acceptable to both parties.

Consultants and clients can work together in terms of fees in many other ways, including performance fees and incentive fees. The important point is that they should be willing to work together on a fixed-fee basis when the scope of the work can be established in advance and when the consultant can reliably estimate the time that will be needed to achieve the desired results.

Finding a Qualified Consultant

Finding qualified consultants requires considerable work by the client. The market for consulting services, the meeting place for consultants and clients, is one of the most imperfect markets in the world. We are used to thinking about perfect markets like the stock market, which is about as close to a perfect market, sans manipulation, that we can find in that it provides instant setting of price and reflection of supply and demand. Consulting is absolutely at the other end of the spectrum. It's a chance occurrence that a consultant and a client ever meet, and therefore both parties have to expend a great deal of effort to ensure that they meet the right party to do business with when the need arises.

The best way to start a search for a consultant is to define your needs and objectives very carefully. What exactly must this person do? What skills must this individual have? What is absolutely essential, and what can you do without? Try to figure out as precisely as possible what capabilities this consultant needs.

Sometimes consultants come to you. You get brochures, you get cold calls. Some consultants do a certain amount of selling. If you have a need at the very moment that somebody who can satisfy that need appears, then obviously your worries are over. But this is very unlikely to occur. Many people start consulting when they are in the right place at the right time, but soon learn that it was just a chance occurrence and do not stay in the business because they are not disciplined or organized marketers.

Undoubtedly, the best way to find the right consultant is through referrals. You need to describe your specific needs and the capabilities you need to people in your company, industry, or business in general, and ask if they know of a consultant who can meet those needs. This approach is sometimes a problem when you are working in a highly proprietary and competitive area. Your competitors may be unwilling to share the names of qualified individuals with you, or you may be unwilling to reveal the nature of your need because it might give your competitors an advantage.

Advertising is another possibility. Consultants read various technical publications, and a very specific advertisement may draw a very good response. Use a box number so that you don't have to reply to everybody. Some consultants advertise in these same publications, and you can usually have an initial telephone interview to determine if one of them can be useful to you.

Still another tool is consultant directories. The consulting profession has members in probably 20 or 30 professional societies, and almost all of them publish a directory of members. The most widely used directory is not published by a professional society, and that is the *Gale Directory of Consultants and Consulting Organizations*, which is in major libraries.

Think creatively. You might contact trade associations, professional associations, executive associations, and publishers and editors of various types of publications, particularly industry newsletters. You may even create a consultant by saying to someone whom you know would be ideal for the job, "I know you aren't a consultant, but would you be willing to tackle this job?"

Finally, in the search for a consultant, you can use a consultant broker. The fee involved in this approach is paid either by the client or the consultant. A certain bias is built in here because brokers are highly unlikely to refer you to someone they don't represent.

Evaluating the Potential Consultant

How do you evaluate a consultant once you have found one? Probably the best initial evaluation comes in the first meeting face to face or by telephone. What is the individual communicating to you; how relevant is it? Do you feel a sense of rapport? Does the consultant seem to have answers for the questions you are asking? Do you feel comfortable in the encounter? By all means, check out references. Any intelligent consultant, however, is certainly going to supply only favorable references, so you need to do an investigative job on your own.

Be a good interviewer and listener. Do not be so eager to dump responsibility that you retain the first consultant with whom you talk. You need to ask about prior experience: With whom has the consultant worked; what problems have been involved; what is the consultant's track record in problem solving? You also need to communicate your own needs, problems, and expectations very clearly. What do you actually require of this consultant so that when the consultation is finished and the requirements met you will be happy.

My final recommendation is that whenever possible you begin the relationship on a short-term basis. Make no long-term commitments until you know one another.

Client Fears

In research I did with more than 600 clients, I have identified certain fears that clients have when dealing with consultants. I list them in priority order:

1. The consultant may look good and sound good but really be incompetent. Satisfy yourself at the initial meeting that the consultant is competent.

2. Worry over whether the client will maintain managerial control of the consultant: Will the consultant run amok in the organization?

3. Is the fee excessive? Justify the fee before you start the consultation.

4. Will the consultant allot enough time to complete the job adequately and in priority order?

5. Does the need for a consultant imply that the client or client organization is inadequate? Make sure that the motivation for hiring a consultant is established without any feelings of guilt.

6. Disclosure of sensitive or secret data: A consultant whose integrity is in doubt should not be retained.
7. The worry that the needs analysis has not been proper or thorough: If the client organization has not done the analysis properly, it should have the consultant do it. But get it done before the job starts.
8. Clients are also concerned that the consultant will foster dependency and that one job will lead naturally into the next. That is not normally the case, and a good consultant should be pointing out the ways that consulting dollars should not be spent—things that should be done internally without need for outside help.

Making the Client–Consultant Relationship Function Smoothly

Do not use a consultant unless you really need one. Consultants will not do a good job unless you give them very clear directions or the responsibility for finding the directions. Be willing to pay what their services are worth. Don't nickel-and-dime on small considerations. If your resources are limited, then limit the scope of work. Consultants charge what they think they have to charge to stay in business. Sometimes they make too little money and sometimes they make too much because they can't always estimate correctly. Neither client nor consultant likes to be taken advantage of, so set measurable objectives—what is to be accomplished and when it is to be completed. Think through your problems beforehand.

Prepare your organization for the arrival of the consultant. Ensure that your employees do not see the consultant as a threat to their job security. Consultants need to be introduced properly and supported and, ideally, need someone in the organization to run interference for them in order to be most effective.

Work only with a written agreement. It need not be a formal contract that only a battery of lawyers can understand, but it should specifically spell out each party's obligations and responsibilities.

Make time to manage the consultant. Consultants do not do well when being supervised, but they do very well when being managed. Maintain open, solid communications. Do not allow the relationship to deteriorate to a point that is not in your best interest.

Things do not always go the way we thought they would, and sometimes the client–consultant relationship gets into trouble. Make sure that the consultant is focusing on creating the best results for you. Pay consultants promptly to keep them happy. Communicate all of your objectives; don't have a hidden agenda. If you cannot formally write out your objectives, you can certainly communicate them verbally. Require regular verbal or written progress reports so that you are continually aware of the progress being made. If possible, assign a staff member with some degree of authority as liaison between the consultant and your organization. Most importantly, give feedback to the consultant and be receptive to feedback from the consultant.

Chapter 4

An Academic Perspective on Consulting

Earl S. Huyser

The key word in the title of this chapter is "an" because this chapter presents only one academic perspective on consulting, namely my own. Since joining the staff of the chemistry department at the University of Kansas in 1959, I have also been actively engaged as a consultant for the Dow Chemical Company. In the course of 29 years, I have spent very little time discussing with others involved in consulting what I do as a consultant. My experience may be somewhat different from that of many other consultants, both in what I do and in how I go about it. It has often been said that one should not try to fix something if it is not broken. Whatever I have been doing as a consultant must be working to some degree or I would not be asked to return year after year. Being invited to contribute this chapter, however, has given me a good reason to examine what I have been doing as a consultant. My account may not reflect in any manner how other academic people go about their consulting for industrial research organizations nor how they perceive what they are doing.

I shall focus on three aspects of my consulting. The first is with whom I consult—my clients—and in what areas of chemistry. The second is what actually happens—what do I do when consulting. And third, I will give some of the reasons that, as an academic

2106–5/91/0041$06.00/0 © 1991 American Chemical Society

person, I have found consulting in an industrial research organization a valuable and rewarding experience.

Clients and Fields of Chemistry

I spend my consulting time with people who do research, and we discuss their research. In the late 1970s when the number of people on the Dow consulting staff was rather limited, one of the research directors asked me to explain what I was doing that kept me on the consulting staff. I told him quite simply that I didn't spend consulting time talking with research administrators. He seemed to understand and apologized for using my time in that capacity. I have made a rule of avoiding spending consulting time with administrators because I have no expertise (nor, for that matter, any real interest) in the areas in which they are working, namely administration. (I do not mind at all going to lunch or dinner with administrators and, in fact, some of them are among my best friends.) With research matters it is very different because I do have experience and an interest in research.

Spending most of my time on research matters influences greatly the people with whom I consult. For the most part, they are the young people who have not yet moved out of research into other areas. Also, I spend much of my consulting time with research associates and research scientists, the people who have made active laboratory research the main thrust of their professional careers. Among the latter are people and projects with which I have been involved continuously for many years. With the young people, consulting is really a great pleasure because they are just becoming involved with a new aspect of their professional careers. They are not only very well informed about the chemistry in their general areas of research, but are enthusiastic about their work and eager to discuss their research with an academic person.

I consult mostly with chemists, but on many occasions I have had very good sessions with chemical engineers. Although we may not always have the same approach to research, I find such consulting sessions very stimulating and, for my part, very informative.

Consulting Activities

I consult with new people in much the same way as I do with those who have had many years of industrial research experience. It is pertinent here that I am hired as a general consultant in organic chemistry. Although I have expertise in some more specific areas of organic chemistry, I do relatively little consulting in these rather narrow fields. Therefore, I do not enter a consulting session as an expert. As a matter of fact, I take a rather different (and realistic) approach and regard my client as the expert in the area under discussion. In many instances, the client may well be one of the world's leading experts in that area. Even if this is not the case, the client has spent far more time directly concerned with the chemistry we are discussing than I have, and it would be presumptuous of me to assume any comparable degree of expertise in that subject.

In the consulting session, the client and I come together as scientific colleagues to discuss in detail some research that is actively being pursued. Sometimes a direct question may require an answer. As often as not, in the detailed discussion of the problem, the client comes up with many good suggestions that lead to the answer. The significant thing about these consulting sessions is that researchers are given the opportunity to discuss in great detail some aspect of their work. While we make every attempt to find answers to specific questions, we also consider the scientific aspects of the work. The latter point is particularly important for a number of reasons. Not the least of them is that research is very hard work, and one of the most important driving forces behind good research is the excitement of the people who are doing it. When research people are excited about their work, experiments will be designed and, more important, may actually be performed in the laboratory. It is very possible for good investigators to become so involved in the more mundane aspects of the work that they lose sight of its scientific merit. Becoming aware again of this aspect of the work can bring back the enthusiasm that is so necessary to successful research.

Another aspect of my consulting results from its longevity. I have been around the Dow research organization a long time and also have had the opportunity to consult at five or six different geographic locations where Dow research is being conducted. Currently, I consult regularly at three of these locations. This experience has given me significant exposure to what is happening in various areas of Dow research. Equally important, I have been exposed to many aspects of Dow research during the past 30 years. Often a situation will come up in a consulting session that reminds me of something that has been investigated before by someone in Dow research. I can then refer the client to a scientist in the company who has had first-hand experience in the area. Dow makes every attempt to maintain effective communication within the organization. Still, work may have been done 15 or 20 years ago by someone whose present research interests give little indication of that earlier experience and expertise. Although communication is very good among people at the different locations of Dow research who are involved in the same areas of work, on several occasions I have been able to direct a researcher at one location to someone at another location who had had experience in a particular area of chemistry at some time in the past.

Rewards of Consulting

Why might an academic person want to be involved with industrial consulting? One reason, of course, is the money. It does pay quite well. The number of days that a professor can make available for the purpose is limited, but my income from consulting was assuredly significant to me when I was an assistant professor in the late 1950s.

Today, however, other reasons are more important to me than the income. Before joining the University of Kansas, I worked as a research chemist at Dow for a couple of years. During that period, I became very interested in industrial chemical research—both how it was done and its effect on society. As a consultant, I have been able to stay in active contact with this part of the chemical profession. Consulting has enabled me not only to stay aware of what is going on in parts of the chemical industry but also to do so by personal contact with the people doing the research.

In time I found another advantage: I could observe what was happening somewhat less subjectively than if I were still actively doing industrial research. I could evaluate more objectively such matters as the impact of industrial chemistry on other areas of the chemical profession, in particular the academic side. We have, for example, seen a dramatic turnaround from interest in process development in the 1960s to interest in new products today. Thus, while physical organic chemistry flourished in academe when process development was important, academic and industrial chemists alike today are far more interested in synthetic chemistry. Similarly, environmental concerns have made analytical chemistry a vital part of all industrial work, and the demand for analytical chemists is reflected in the increase in graduate enrollments in this area.

As an academic who has the advantage of a significant amount of contact with industrial research, I have attempted to act as a bridge between these parts of our profession. In general, I have found that industrial people are far more interested in developments in academe than are academics in what is happening in industrial research. This may be because everyone in industry has had direct contact with academe and so is acquainted with the academic world to some extent, whereas the opposite is not always true. Many academics have had, at best, only limited association with industrial research. One result, unfortunately, is that many academics have developed erroneous conceptions about industrial research, often lack interest in it, and, in some cases, do not appreciate it.

This is not the case, however, with most students. I have found that graduate and undergraduate students alike develop great interest in professional opportunities in industrial laboratories. This interest seems to become most intense as they near graduation and are about to embark on careers in industrial laboratories. One of the most satisfying aspects of my professional life in teaching and research has been to be able to communicate the excitement of industrial research to these young people at this stage of their professional careers. Without having had the opportunity of being a consultant and developing professional relationships with many industrial scientists, I would not be able to tell students of the professional opportunities they are about to encounter.

My professional experience as a chemist has been very satisfying in that I have been able not only to teach and do academic research, but also to stay actively involved, through consulting, in some aspects of industrial research. Certainly, one of the noteworthy rewards of my active contact with people in industrial research has been its direct and positive effect on what I am doing as a member of the academic community.

Chapter 5

University–Industrial Relationships

Charles S. Sodano

Relationships of private businesses and universities are idealized by many as multipurpose arrangements: the businesses put up cash and receive benefits that include a source of new employees who are trained, somewhat, to their specifications, as well as technical information.

It is, of course, very difficult to dictate specific curricula that will serve a company's interest. Mechanisms are available, however, for steering the educational process in the desired direction. They include the hiring of interns, purchasing equipment or facilities for a university, and other approaches that increase the understanding of the educational elements that will get a student a professionally and financially satisfying job with the company.

In addition to a source of employees, industrialists expect universities in these arrangements to supply technical information packaged in a variety of ways. They include:

1. educational seminars and training programs to keep the industrial staff up to date and versatile

2. research programs that provide information that the in-house staff cannot effectively provide to support long-range business plans

3. assistance in solving short-term technical problems

Help with short-term problems is the one perhaps the least utilized by companies, but it can help to fill a manpower gap that exists in many of them.

It is not usually the role of universities to provide routine repetitive services for industry. Such services do not ordinarily serve the charter of academic institutions to educate and train students. Moreover, universities should not be in competition with the private laboratories that provide these kinds of services.

The technical staffs of many chemical businesses are much smaller than was the norm several years ago. These staffs normally have two major functions. The first is to deal with issues spelled out in the company's annual business plan. These issues often involve short-term financial gains. The second major function is long-term research that could lead to profits. These research programs are normally reviewed at least annually to determine whether their cost is justified.

During the normal business year many problems arise that require the immediate attention of the technical staff. Examples include manufacturing difficulties, possible violations of government regulations, and consumer complaints. There may also be a long list of technical feasibility studies that should have been explored by in-house staff. Without outside assistance, or staff expansion, the company's technical staff must postpone some current or long-range task to handle these problems. Often it would be feasible to get assistance from a university. Normally, however, this is difficult to do on short notice unless appropriate university–industry relationship exists.

Many organizational arrangements can be set up between a university and a company. Fundamentally, the relationship can be one-to-one or one-to-many. The many is usually the industrial clients, but it could be a pool of several universities. Typical popular arrangements include:

1. an outright gift to a university to be used for activities of interest to industry

2. individual consulting or contractual arrangements with a professor at a daily, annual, or job rate

3. pooled fund arrangements where members contribute annual dues and can influence the course and selection of research projects. The mean going rate for dues in these organizations is about $30,000 per year.
4. informal or formal meetings with university staff to discuss mutual needs and wants
5. pooled fund arrangements where each industrial member contributes about $15,000 annually and gets short-term problem-solving (STERMPS) assistance plus individual consulting, open discussions, and a chance to influence the training of students

Featuring help with short-term problems is the focus of this chapter.

STERMPS: What Is It?

STERMPS encompasses a variety of tasks. Examples are:
1. brainstorming sessions where ways to solve technical problems are suggested and evaluated by a team of university and industrial scientists
2. use of specialized equipment to perform a critical analysis that the company cannot handle. This task includes provision of the equipment and expertise in running it and interpreting the data.
3. short-term developmental research, normally entailing a professor's applying known technology to a commercial material
4. feasibility studies
5. conducting training sessions to transfer new or improved technology
6. review and evaluation of technical literature

How To Start a STERMPS Program

What the University Has To Do. The key feature of a STERMPS program is the university program director. It is essential that program directors spend the majority of their time acquiring and finishing projects effectively. To provide a focal area, the university program should be established within a formal center of expertise that encompasses several disciplines. This arrangement does not preclude the participation of university staff who are not part of the center.

The starting point for soliciting industrial associates for such a program would be companies who already have some interest or activities in the center. To make the program manageable, the number of industrial associates should not exceed 15. The annual membership cost per company should be in the range of $15,000. This is enough money to keep everyone interested, but not enough to create grandious expectations of funded projects.

The program director functions as an industrial project leader within the university in planning and meeting project goals. Participating university staff receive per diem or other compensatory funds for their efforts. Participating staff must meet project deadlines;those who do not meet deadlines should not participate in the program. The program director may have to nag (as an industrial project leader often does!) to ensure that deadlines are met.

In addition to staying on top of active projects, the program director must maintain current knowledge of the expertise and capabilities of the university staff. This kind of information is supplied to the industrial coordinators so as manner to keep them aware of the university staff's capabilities and to help them select those skills and personalities that match their projects.

The program director must keep projects small enough to be funded with the resources available. Some projects can be split into a feasibility study funded by the program and further work funded by additional support from the sponsoring company. Projects that cannot quickly be brought to some kind of conclusion should not be included in the program.

Meetings should be held periodically for all industrial coordinators to introduce them to university activities that potentially could be useful to their companies. Such meetings also foster the development of intercompany relationships.

What the Industrial Associate Has To Do. The industrial coordinators in STERMPS programs have coresponsibility for project management. They must have an active grasp of their companies' short-term problem-solving needs and match them with university resources. Their role in project management is to link their technical associates with these resources and, in effect, use the university people as adjunct staff. Industrial coordinators should evaluate projects to determine whether they can be funded by the program or must be handled separately. This decision usually requires some discussion with the university program director.

The coordinator also must keep the industrial technical staff aware of the university's skills and capabilities. This is done by inviting faculty to give seminars on topics of mutual interest or to participate in informal meetings to explore activities potentially of interest. Periodically, the coordinator should periodically contact the university program director to brainstorm new areas of mutual interest.

Benefits to the University. The university in a STERMPS program, after expenses, could have excess annual dues funds in the neighborhood of $50,000. These funds could be applied to the program, or to any other program, according to the university's needs. Projects often expand beyond the means of the program. Such projects require separate funding, and some evolve into multiyear efforts. The university also has a select audience to showcase faculty, facilities, or anticipated expansions.

The key benefit of the program, however, is the development of personal relationships between the industrial and university staffs. These relationships ultimately benefits both university and companies beyond the scope of the program. The resulting appreciation of professional skills, information networks, and personal introductions enable the participants to freely utilize resources that would not normally be readily available to them.

The university staff has a unique opportunity to share in the solving of industrial problems. This experience gives the staff an appreciation of what is important to running and maintaining a business and what areas of long-term research could be useful to these associates.

The university students can benefit directly by performing tasks in a project management style that many of them will encounter when they take industrial positions, perhaps at one of the client companies. They also benefit indirectly because some research projects that evolve from the program will develop information that could be directly used by companies to develop new or improved products or processes. This, of course, makes the students more desirable future employees.

Benefits to the Industrial Participants. The immediate benefit of a STERMPS program to a company is, of course, the equivalent of an increased technical staff. Companies also gain the opportunity to use many university resources that would not usually be known or accessible to them. The indirect cost savings could be substantial. The cost of some of the projects, if handled internally or at nonuniversity centers, would usually be substantially greater than under the STERMPS program.

The industrial staff also develops an independent university consulting staff that is very familiar with its problems and operations and therefore is able to react quickly. Finally, participating companies have the opportunity to wield some kind of influence in the training of students so that their skills more closely match the company's needs.

Chapter 6

Accessing Federal Laboratories Know-How

Lee W. Rivers

Our nation's federal laboratories are grossly underutilized by American industry. That's a fact, not an opinion. We have a unique opportunity as a nation to correct that situation, but it will take hard work by industry and government. Is it worth the effort? What are the potential benefits for the chemical industry consultant? Why is it so important that our nation's federal laboratories be used more by American industry than is the current practice?

First of all, the nation has a very serious competitiveness battle on its hands. It's not a new problem. It has been festering for 40 years and boiling over for the past 8 to 10 years. In 1980 the United States had a $27 billion positive trade balance with the rest of the world in high technology products. In 1986, the high-technology trade balance went negative! Even with the change in exchange rates the dollar experienced in 1987, the high-technology trade balance barely returned to a positive level.

One after another of the nation's industries is suffering tremendous inroads from foreign competition. In 1970, the American domestic consumer electronics market was supplied entirely by American producers. Today, they supply less than 5% of our

2106–5/91/0053$06.00/0 © 1991 American Chemical Society

domestic market. Our educational system, kindergarten through 12th grade, is underperforming the educational systems of most other nations. The results of testing of our young people show them consistently scoring lower in science and mathematics than young people from other industrialized nations and from many of the newly industrialized countries. It is estimated that by the year 2010 a shortage of 500,000 scientists and engineers could exist in this country. How can we compete globally without a skilled professional and technical work force? These developments are only part of the litany of what has happened over an extended period to weaken our competiveness in the global marketplace.

Some economists say, "Let everything equalize in the world marketplace and this country will end up doing what it does best. If we wind up making Mrs. Fields' Cookies for the rest of the world, that's all right." Don't you believe it! I don't believe it for a moment. I think that the standard of living for any worker is derived in large part from the value to the product or service. I don't believe that making Mrs. Fields' Cookies is going to earn future generations the standard of living enjoyed by those people who are making the ceramic engines and high technology drive trains for the car of tomorrow. I don't think we will have as high a standard of living if all we do is assemble the car and paint it while we import all of the high technology components. I think it is important for the standard of living of future generations that we have high-value added jobs in this country.

We generate science at a clip that outpaces all other nations, Japan and Germany included. We spend about $15 billion a year pumping knowledge into the world's reservoir of science and technology. We have the world's best research university system that generates knowledge at a fantastic rate—10 times the rest of the world.

The problem is that we, as a nation, no longer control the rate at which that scientific knowledge is converted into products, goods, and services for the world marketplace. At one time, we were the world's pacesetter for that conversion. The paradox is that we're probably faster at the conversion than we were 10 years ago. If it took us 10 years a decade ago to convert science into a product, we do it today in 7 or 8 years. If we've picked up the pace, what's the problem? The problem is that our international competitors have found a way to convert science, often our science, into products, goods, and services in three years or less. So although we may have improved over time, we can't be complacent, because many of our international competitors are doing even better than we are at the conversion process. With that as background, where do the federal laboratories come into play?

Role of Federal Laboratories

The United States is said to have 700 federal laboratories, but we don't really know how many we actually have. What is a federal laboratory, what isn't a Federal laboratory? Typical of Washington, we don't know when to stop counting. So we say we have about 700. The reason the number is wrong is interesting, but the reason the number is significant is more important. That's a lot of federal laboratories, whether it's 620 or 710. Most people, if asked, would guess that we have about 25 federal laboratories, but the number is much closer to 700.

We say we have 100,000 scientists and engineers working in our federal laboratories. We don't know that it is 100,000, so that number is wrong, too. But whether its 97,000 or 108,000, the significance, again, is that it's a big number. One-sixth of the nation's total pool of research scientists and engineers work in those federal laboratories.

We say the nation spends $20 billion annually on research, development, testing, and evaluation in its federal laboratories. That number is wrong, too, but whether its $18 billion or $21 billion, remember that its a big number!

What do the numbers 700, 100,000, 20 billion tell us? They tell us that we have a gigantic resource in our country. What does our history tell us about the utilization of that resource in helping American industry to compete in the global marketplace? Let's look.

The federal government spends more than $60 billion annually on research and development (R&D) and so does private industry. As a nation we spend $120 to 130 billion. That means we spend one-third of the federal government's R&D dollars or one-sixth of the nation's total in our federal laboratories. It's an enormous part of our total national capability.

Enabling Legislation

Until 1980, by design, the federal research establishment was kept fairly isolated and out of the picture in terms of active, constructive engagement with American industry. The prevailing theory was that both are doing their jobs—let's leave them alone.

In 1980, Congress started to pass legislation that affects federal research. That's the year it passed the Stevenson–Wydler Act and the Bayh–Dole Act. Both acts began to open the federal laboratories—first to universities, small businesses, and nonprofit organizations. The legislation extended certain intellectual property rights to those groups. Why did Congress do this? It took the position that jobs were really the issue. You can reach members of congress when you talk to them about jobs. That's one thing they really understand. How are the constituents in their districts doing?

It was obvious that high-technology industries were springing up around our research universities—Route 128 in Massachusetts and Silicon Valley in California. Why couldn't the same thing occur around federal laboratories? Certainly among the 100,000 scientists and engineers working in those laboratories are some excellent ones; eight Nobel Prizes over the years to scientists in just one federal laboratory attest to that.

So Congress in 1980 started the process. Then came changes in patent procedures. The President issued a Patent Memorandum in 1983, and Congress passed another piece of legislation affecting patent matters in 1984. Congress blew the doors wide open with the Technology Transfer Act of 1986. On April 10, 1987, President Reagan signed Executive Order 12591, which gave marching instructions to the agencies of the Federal Government to get on with making federal R&D accessible to American industry. It was very important for the Administration to embrace this legislation and in simple terms say to American industry "come and get it". During a conference on superconductivity held in Washington in July 1987, President Reagan gave a forceful speech on technology transfer. Superconductivity happened to be the subject, but the speech dealt with making our federal laboratories accessible. The President said in part, "The message of Government is simple—we have an open door policy to the private sector; cooperation, wherever and whenever possible, is the order of the day".

What's the problem? We have the enabling legislation. We

have the President's Executive Order. Why isn't industry using the federal laboratories more effectively?

Barriers to Accessibility

First of all, there is a bureaucratic reason. After a law is passed by Congress and signed by the President it takes the bureaucracy, on average, two years to put in place the rules and regulations needed to implement that law. So although Congress put good incentives in the law to encourage the laboratories to work with American industry—royalties from licensed inventions now flow directly to the federal laboratory (not the U.S. Department of the Treasury) and 15% or more of the royalty flow to the laboratory goes to the inventor—implementation has been slow. Despite slow progress, Congress and the Administration remain dedicated to seeing that American industry and the federal laboratories work together more effectively, for the common good.

What about American business managers? What is their attitude toward working with the Federal Government and specifically with federal laboratories?

If an American business manager visits a federal laboratory and finds science or technology being worked on which is of interest to the manager, and can make a deal right there at the laboratory, the manager may do it. The average business manager does not want to deal with the Federal Government bureaucracy. It takes too much time and energy to achieve anything.

Therefore, although an impasse is often reached, we can't let it stop there. We must find a way to bring this gigantic federal research system into constructive engagement with American industry. It's too big to omit from the nation's competitiveness battles.

We have incentives. We have a willingness at the federal laboratory level to cooperate with industry. We also have the Federal Laboratory Consortium for Technology Transfer, the FLC.

The FLC—Aid to Accessibility

The FLC began in 1974 among a group of federal laboratory employees who were reaching out to their local communities—primarily their universities and their state and local governments.

Competitiveness of American industry wasn't perceived as the problem in those days, but these scientists realized that their federal laboratories had knowledge that might help the local sewer plant when it clogged up. So they were reaching out. They started to network with one another across federal agencies, and the FLC was born. It grew and thrived on volunteerism. The members wrote their own charter and elected their own chairperson, and they still do that today. The FLC is unique in federal government circles.

The FLC is not a federal agency—it's a network cutting across 14 federal agencies. From 1974 to 1987 the consortium existed on handouts from the laboratories and the agencies who sensed that what these technology transfer networkers were trying to do was important. Therefore, they provided voluntary funds for the work. Since 1987, the FLC has been funded by a levy on intramural spending of each federal agency engaged in R&D. In fiscal 1989, the FLC had a budget of approximately $1.2 million to use to leverage the $20 billion of R&D spending in federal laboratories to benefit the private sector.

The FLC divides the country into six geographic regions, each with a regional coordinator. It operates a computerized data base and a centralized clearinghouse for information about who's doing what in the federal R&D establishment. The clearinghouse is supported by an electronic mail system that cuts across the 14 federal agencies that operate laboratories. Several hundred technology-transfer people assigned to individual federal laboratories are industry' s primary entry point to the laboratories. Get to know these technology-transfer agents and use them and their network to serve your clients. Be a bridge-builder between the federal laboratories and your client in the private sector.

The FLC can identify, at 700 federal laboratories, the people and projects of specific interest to you and your client. It can give you access to the federal laboratories to ferret out the specific science and technology you and your clients need to remain technologically competent and world-class competitors.

When you come to the FLC with your rifle-shot question, what happens? First, the data base is searched. Then the electronic mail system swings into action, and the FLC tells you, for example, "The two scientists who are doing work of interest to you are at Beltsville, Maryland, and the Idaho National Engineering Laboratory." It tells you their names and how it will act as your ombudsman.

FLC representatives understand the problems of dealing with

the Federal Government. Each federal agency is at a different stage in establishing the rules and regulations required to implement the enabling legislation. Also, we know that some of those 100,000 scientists and engineers haven't gotten the message that we really want the laboratories to work with American industry. Therefore, you're going to run into some problems and obstacles, but we hope, not too many serious ones. If you do, go back to the FLC point of contact, the regional coordinator, the chairman or vice-chairman of the consortium, or the FLC's Washington office. We're here to help and support you—and at no charge. Some FLC points of contact are

FLC Chairman
Dr. Loren C. Schmid
Pacific Northwest Laboratory
P.O. Box 999
Richland, WA 99352
(509) 375-2559

FLC Vice-Chairman
Ms. Margaret McNamara
Naval Underwater Systems Center
Code 105, Building 80T
New London, CT 06320
(203) 440-4590

FLC Administrator
Mr. George Linsteadt
Delabarre & Associates, Inc.
P.O. Box 545
Sequim, WA 98382
(206) 683-1005

FLC Clearinghouse
Mr. Allan Sjoholm
Delabarre & Associates, Inc.
P.O. Box 545
Sequim, WA 98382
(206) 683-1005

FLC Washington, D.C. Representative
Dr. Beverly Berger
Federal Laboratory Consortium
1550 M Street, NW, 11th Floor
Washington, DC 20005
(202) 331–4220

Midwest Regional Coordinator
Ms. F. Sinclair Ingalls
Wright Laboratory
WRDC/XO
Wright–Patterson AFB, OH 45433–6523
(513) 255–2006 or 8997

Northeast Regional Coordinator
Mr. Al Lupinetti
DOT-FAA Technical Center
ATTN: ACL-1 Atlantic City International Airport
Atlantic City, NJ 08405
(609) 484–6689

Midcontinent Regional Coordinator
Mr. Arthur Norris
National Center for Toxicological Research
County Road #3 (HFT-2)
Jefferson, AR 72079–9502
(501) 541–4516

Southeast Regional Coordinator
Mr. Robert Barlow
John C. Stennis Space Center
Code HA30
Stennis Space Center, MS 39529
(601) 688–2042

Midatlantic Regional Coordinator
Mr. Nick Montanarelli
SDIO
SDIO/T/TNO
Washington, DC 20301–7100
(202) 653–1442

Far West Regional Coordinator
Ms. Diana Jackson
Naval Ocean System Center
Code 014
San Diego, CA 92152
(619) 553–2101

CONSULTING OPPORTUNITIES

Chapter 7

What Consulting Practices Look Like

The Nuts and Bolts of Organizing a Practice

Michael Curry

The consulting sector of the chemical and allied industries is highly varied and difficult to categorize. Many consultant firms are large and have most of the functions of operating companies. At the other extreme are many individual consultants who work out of an office in the home, sending out reports generated on a personal computer with data obtained from the literature and telephone calls. To roll such diverse consultancies into discrete, homogeneous sets is impossible.

An approach that might shed some light, however, would be to assume that an association of consultants might have a commonality simply because they have associated. Such a group is the Association of Consulting Chemists and Chemical Engineers, based in New York City has existed for 60 years. It requires that its members join as individuals, although they may be part of a larger organization. The association's strongest asset is probably its computerized data base of members' expertise, which is accessible to potential clients through a call to the office. Potential clients also can purchase a companion directory that lists the same data.

The association meets monthly for a business meeting, a social hour, dinner, and a prepared lecture. It publishes a quarterly newsletter about consultants with an audience of 2000.

Fifty percent of the members have belonged for more than four years. The association has been the same size—100 to 125 consultants— for a decade; about five leave each year for a variety of reasons. Many of those who join are new to consulting, and belonging to an association of peers is important to their orientation to consulting practices as well as for personal contacts. An analysis of such an association, then, would be based on a better than randomized cross section of consultants.

Every other year, the association surveys its members and collates and mails the results to them for their own use. Following are the data from the last survey with comments on the replies.

INDEX

I) Backgrounds of Respondents

II) Organizations
  1. Size
  2. Degree Levels
  3. Use of Other Consultants
  4. Type of Consulting
  5. Facilities
  6. Consulting Areas

III) Clients
  1. Number and Length of Association
  2. Types
  3. Major Contacts
  4. Size of Clients
  5. Major Client Fields

IV) Fee Structures
1. Per Diem Basis
2. Fees for Lab Time
3. Fees for Non-Lab Time
4. Minimum Billing
5. Expense and Billing

V) Business Development
1. Major Sources
2. Most Important
3. Promotional costs
4. Benefits

Association of Consulting Chemists and Chemical Engineers Data on Questionnaire on Consulting Practices
January, 1987

I. BACKGROUND OF RESPONDENTS

1. Years of Consulting:

1 year -	17	9 years -	2	26 years -	2
2 years -	5	10 years -	3	30 years -	1
3 years -	7	11 years -	3	34 years -	1
4 years -	2	12 years -	1	38 years -	1
5 years -	7	17 years -	1	39 years -	1
6 years -	1	19 years -	1	41 years -	1
7 years -	2	20 years -	1		
8 years -	2	25 years -	1		

2. Type of Background

Chemist	Chemical Engineer	Other	Total
36	25	2	63
		(Marketing, Metallurgy)	

II. ORGANIZATION

1. Size of Professional Staff

1	2 - 5	over 5
45	14	8

Curry *What Consulting Practices Look Like*   69

2. *Highest Degree Levels (first five members)*

                                                                                   Total

Staff of One	D*(22)	M(9)	B(14)		45
Staff of Two	DD(1)	DM(1)	DB(1)	M-(1)	6
Staff of Three	DDD(1)	MMB(2)	MBB(1)	BB(1)	4
Staff of Four	DDDD(1)	DDMM(1)	DBB-(1)	MB(1)	4
Staff of Five or more	DDDDD(1) DDBBB(1)	DDDMM(1) DMMMM(1)	DDMMB(2) DMMB-(1)	BBBB(1) DDMBB(1).	8

* D = Doctorate Degree, M = Masters, B = Bachelor's

Comments: 40 of 67 are one-consultant operations. 50% of one-person operations are PhD's.

3. *Use of 'Ad Hoc' Consultants*

                               Used - 36   Not Used - 2   Total - 38

Is Client made aware of the use of 'Ad Hoc' Consultant?

                           Yes - 17   No - 1   Sometimes - 1   Total - 19

Comments: The use of ad hoc consultants appears to be universal.

4. *Types of Consulting*

Mostly lab work-3   Lab and Personal-26   Mostly Personal-39   Total-68

Comments: The largest number do little or no lab work. Only three do "mostly lab work." These results are expected.

5. *Facilities Used*

a) Laboratories

		total
1) Principally own facilities		18
2) Outside facilities		30
3) Other		4
		52

b) Offices for Personal Consulting

1) Owned or Rented	31
2) Home	29
3) Clients	-
4) Other	8
(Includes combinations)	68 total

Comments: It is interesting to see how many do lab work principally in "out of house" facilities, and almost one-half of all nonlaboratory consultancies work out of their homes.

6. *Major Areas of Consulting (10% or more of total billings)*

1) General Management 4) Manufacturing 7) Other
2) Technical   5) Engineering
3) Sales/Marketing   6) Planning

Responses by Category

One Response	#2(15)	#5(3)	#7(1)
Two Responses	#2,5(5)	#2,4(4)	#2,7(3)
	#2,3(2)	#2,6(2)	#3,4(2)	Subtotal 19
Three Responses	#2,3,6(3)	#1,2,3(2)	#1,2,6(2)	Subtotal 18
	#2,4,5(2)	#2,4,6(2)	-7 different threes	Subtotal 18

Four Responses   Subtotal 7
Five Responses   Subtotal 4
Six Responses   Subtotal 1
                                     67

Other included: Failure Analysis, Environmental, Expert Witness, Regulatory Compliance, Seminars, Acquisitions and Divestitures. Due Diligence, Arbitration and Litigation, Mergers, Custom Processing, Risk Evaluation, and Patents.

Comments: The data are scattered with 1, 2, and 3 responses of equal size. Technical was the predominant response.

For which of the above is the consultancy best known?

(Technical)	35	(Sales/Marketing)	5
(Engineering)	7	(General Management)	1
(Other)	5	(Planning)	1

Total - 54

III. CLIENTS

1. *Number of Major Clients (5 days or more) in 1988*

   1 - 5 =(29)   6 - 10 =(18)   >10 =(16)   Total = 63

2. *Percentage of Major Clients who have been clients 2 years or more*

	0%	1%	10%	20%	30%	33%	40%	50%	60%	66%	75%	80%	90%	100%
Number:	6	1	2	4	3	1	1	4	7	1	3	5	3	11

3. *Types of Clients*

   1) Operating Companies 3) Consultants 5) Other
   2) Governmental Bodies 4) Lawyers

   Responses by Category

Single Response	#1(20)		Subtotal=22
Two Responses	#1,4(11)	#3(2)	Subtotal=25
	#1,2,4(6)	#1,2(6)	
Three Responses	#1,2,3(1)	#1,3,4(5)	#1,3(6)	Subtotal=16
		#1,2,5(1)	#1,3,5(2)	#1,5(2)
Four Responses			#2,3,4(1)	Subtotal= 2
				65

   Comments: In all categories, operating companies dominate; lawyers, government bodies, and other consultants are high.

4. *Position of Person who is the Major Contact in the Client*

   1) Chief Officer or Owner   5) Marketing Manager
   2) Other Officer   6) Engineering Manager
   3) Technical Manager   7) Other
   4) Manufacturing Manager

   Responses by Category

Single Responses	#3(9)	#1(7)	#2(3)	#4(1)	Subtotal=20
Two Responses	#1,3(7)	#1,2(3)	#3,7(2)	#2,3(2)
	#3,4(2)			#3 different twos

Three Responses	#1,3,5(4)		#1,2,3(4) #2,3,5(2)
	-6 different threes		
Four Responses			Subtotal= 7
Five Responses			Subtotal= 2
Six Responses			Subtotal= 2
			66

Comments: The technical manager dominates as a contact. "Other" was a series of different people such as lawyers (4), division managers, entrepreneurs, quality control managers, law firm partners, research and development, new business development managers.

5. *Size of Major Clients*

1) Under $1MM sales   3) $51-1,000MM sales
2) $1-50MM sales   4) Over $1 billion sales

Response by Category

Single Response	#2(13)	#4(5)	#3(4)	#1(3) Subtotal=25
Two Responses	#3,4(6)	#2,3(4)	#1,4(3)	Subtotal=17
	#2,4(2)	#1,2(2)		
Three Responses	#2,3,4(10)	#1,2,4(2)	#1,3,4(1)	Subtotal=14
	#1,2,3(1)			
Four Responses	#1,2,3,4(8)			Subtotal= 8
				64

6. *Major Fields of Clients*

1) Health Care   5) Minerals, Metals   9) Consulting
2) Cosmetics   6) Polymers, Plastics   10) Energy,
3) Food   7) Specialties, Fine   Environment
4) Agricultural   Chemicals   11) Other
   Chemicals   8) Engineering

74   TRENDS IN CHEMICAL CONSULTING

Responses by Category

Single Response	#6(10)	#10(2)	-7 different ones	Subtotal=19
Two Responses	#6,7(4)	#7,11(2)	-9 different twos	Subtotal=15
Three Responses	#6,7,11(2)		-13 different threes	Subtotal=15
Four Responses	#1,6,7,11(2)		- 6 different fours	Subtotal=8
Five Responses				Subtotal=3
Six Responses				Subtotal= 1
Seven or More Responses				Subtotal= 3

Comments: Write-ins for #11 included industrial chemicals (2), petroleum refining, surfactant manufacturing, computers, fibers, legal, seeds, electronics, mechanical, law firms (2), expert witness, paper, heavy inorganics, OEM (distillation equipment), utilities. pharmaceuticals. The major fields are extraordinarily broad!

IV. FEE STRUCTURE

1. *Per Diem Fees Based On:*

Salary Only - 29   Salary and Overhead - 18   Other - 16   Total - 63

"Other" includes salary and expenses (4), three-times salary, sliding fee scale, fixed fee, hourly rate, what is appropriate, lump sum and bonus, what client will pay.

Comments: Most salary questions were apparently not clear. "Per diem" was intended to be the fee without overhead. The "other" answers, such as "what client will pay" and "bonus," were interesting.

2.
Fee Structure
Fees for Laboratory Time

| Per Diem || Contract ||
Maximum	Minimum	Maximum	Minimum
1400	440		
1100	850	850	550
1000	400	500	400
1000			
1000			
900	600		
800	1200		
800	600	600	400
800	400	800	200
750	600	600	500
700	700	600	600
700	700		
650	550		
640	640	600	600
640		320	
600	500	600	420
600	400		
600			
500	400	500	400
500	400	350	250
500		400	
500			

| Per Diem || Contract ||
Maximum	Minimum	Maximum	Minimum
150	40		
54	30	54	25
		50000	5000

3.

| At Office |||| Away From Office ||||
| Per Diem || Contract || Per Diem || Contract ||
Maximum	Minimum	Maximum	Minimum	Maximum	Minimum	Maximum	Minimum
2000	100			2000	100		
1600	1500	1500	1100	1600	1500	1500	1000
1400	440			1400	440		
1250	1000	1250	800	1500	1250	1500	1250
1100	500			1100	500		
1000	600	1000	500	1200	1000		
1000	300	1000	300	1200	400	1200	400
1000		1000		1000		600	
1000	600	800	500				
1000	600	1000	500	1000	600	1000	500
1000	800	800	600	1000	800	800	600
900	750	700	500	900	750	700	500
900				900			
800	600	800	400	800	600	800	400
800	600			700	200		
800	650			800	650		
800	800	400	400	800	800	800	200
800	400	800	200	800	400	800	200
800	400	600	400	600	400	600	400

	At Office			Away From Office			
Per Diem		Contract		Per Diem		Contract	
Maximum	Minimum	Maximum	Minimum	Maximum	Minimum	Maximum	Minimum
800				800			
800		600		800		600	
800	800			800	800		
800	450			800	450		
800	400			800	400		
750				1000		750	
700	700			700	700		
700	600	700	300	700	600	700	600
650				650			
650				650			
600		500		600		500	
600	500	600	500	600	500	600	500
600	500	600	500	600	500		
600				600			
600				600			
600				600			
600	500	400	400	600	500	400	400
600	500	600	420	600	500	600	420
550		550					
550				650			
500		400		500		400	
500	500	500	500	500	500	500	500
500	400	400	200	500	400	400	200
500	500	480	480	500	500	480	480
500				500			
500	400	400		500	400	400	
500	420			500	500		
500	300	500	300	1100	850	850	550
500	400			700	500		
450	350	550	350				

| At Office |||| Away From Office ||||
| Per Diem || Contract || Per Diem || Contract ||
Maximum	Minimum	Maximum	Minimum	Maximum	Minimum	Maximum	Minimum
400	400	500	400	500	500	500	500
400	300			400			
400		400	350	400	400	400	350
360	360	360	360	450	450	450	450
300				400			
300	200	300	200	500	500	500	500
280		400		500	250	400	250
200	200	200	200	350	450	350	450
150	40			150	40		
100				100			
92	35	92	30	92	35	92	30
		2000	250				
				1500	1000	900	700
				1000	1000	3000	1000
		250	250			250	250

Comments: Because of the scatter, all the data are collated on the basis of the per diem maximum. Ranges are broad: from $2,000 to $30 per day. The median maximum was $650 per day. As expected, per diem fees are generally higher than contract fees.

4. *Minimum Billing*

    1) Under 1/2 day - 17   3) 1 day - 14
    2) 1/2 day - 20   4) Other - 4

                 Total = 55

Other includes: contract minimum, $75, 1 hour.

5. *Billing of Expenses*

  a) 1) Always billed   54   Total = 65
     2) Sometimes billed 10
     3) Never billed   1

  b) 1) Actual Out-of-Pocket 45   Total = 54
     2) Actual - Overhead   7
     3) Other   2

  c) Billings Include:
     1) Phone   3) Office Supplies   5) Other
     2) Travel Time 4) Office Services

Responses by Category

Single Response	#5(5)	
Two Responses	#1,2(7)	#2(3)
	#1,5(3)	#2,5(5)
		#4,5(2)
Three Responses	#1,2,5(11)	#1,2,4(5)

#1(1)	Subtotal= 9
#1,4(3)	Subtotal=21
#2,4(1)	
#1,3,4(2)	Subtotal=21

Four Responses   Subtotal= 6
Five Responses   Subtotal= 5
                               62

Comments: No. 5 includes outside office services, travel expenses, secretarial expenses, expenses, printing, computer time, travel expenses, overhead, hardware, living expenses, lab fees, materials, freight, car expenses, mail publications, subcontracts, equipment rental, photography costs, and documents. The questions apparently were not broad enough to cover all items.

V. BUSINESS DEVELOPMENT

1. *Major Sources of Business Development*

   1) Personal Contacts   4) Organizational Contacts
   2) Advertising   5) Brochures
   3) Referrals   6) Other

Responses by Category

Single Response	#1(8)	
Two Responses	#1,3(18)	
Three Responses	#3(2)	
	#1,4(3)	
	#6(1)	Subtotal=11
Four Responses	#1,2,3,4(4)	Subtotal= 7
Five and Six Responses	#1,3,4,5(3)	Subtotal= 4
		65

Other included representatives, mailings, books and seminars, and publications.

2. Single Most Important is ( ) Above

   #1(37) #3(19) #6(3) #4(2) Total = 61

   Comments: Personal contacts and referrals dominate!

3. Percentage of Gross Billings Spent for Promotion in 1986

0% -	15	5% -	8	15% -	1
2% -	6	7% -	1	20% -	3
3% -	4	10% -	4	21% -	1

   Comments: Largest number spent nothing; four spent 20% (?)

4. Miscellaneous

   1) We do ____ /do not ____ have a pension plan.
   2) We do ____ /do not ____ have professional liability insurance.
   3) We do ____ /do not ____ have medical and hospital coverage.

   Replies

   1) Do - 38   Do Not - 28
   2) Do - 21   Do Not - 45   Total = 66
   3) Do - 41   Do Not - 25

Pattern of Replies   A = DO   B = DO NOT

AAA - 9	ABA - 20	BAA - 7	
BBB - 12	AAB - 3	BAB - 2	Total = 66
	ABB - 6	BBA - 7	

Comments: Less than one-half have pension plans. Only one-third have liability insurance. Two-thirds have medical insurance. The major patterns are:

1. pension and medical but no liability (ABA)
2. none of the benefits (BBB)
3. all of the benefits (AAA)
4. liability and medical but no pension (BAA) 5) medical only (BBA)

Chapter 8

Defining and Marketing Your Consulting Specialty

Peter R. Lantos

Hardly a day goes by that you do not see an article, read a book or book review, or talk to someone extolling the rewards of consulting. But is it really for you?

Consulting is providing a specialized service for a fee. The service can consist of advice or information, but often it extends to doing commercial development for a client, helping to modify a machine, or otherwise implementing the initial findings of an assignment.

The three fundamental categories of consultants are the large firms, such as Arthur D. Little or SRI International; the medium-sized companies, such as Kline & Company, Inc. or Skeist Laboratories; and the small companies, often consisting of one person, such as the Target Group, Cross Gates Consultants, or Kossoff Associates. This chapter is aimed mainly at those who are considering consulting on a small scale, at least initially.

As a prospective consultant, you can offer services in many fields: You can be a technical problem solver, do market research studies, serve as an expert witness, offer contract research, assist with planning, or do analytical testing. A critical point is that you

2106–5/91/0083$06.00/0 © 1991 American Chemical Society

define exactly what you propose to do as a consultant. Although you may consider yourself knowledgeable in a number of fields, and may be inclined to offer your services in all of them, it is very much to your advantage to define a specific field in which you propose to consult. For example, you will be more credible when you claim expertise in olefin polymerization, rather than in all polymerization processes or, even worse, in all chemical processes.

Advantages and Disadvantages

You would be well advised to examine whether you are really intended to be a consultant. The field offers a number of advantages that seem very attractive. You have almost complete independence, you can choose the assignments that really take advantage of your expertise, and, therefore, you can make a real contribution. Prestige is often associated with being the expert; you meet many interesting, although often demanding and challenging people; and you may need to travel extensively.

However, there are some disadvantages: You need much self discipline, as you will be working long, hard hours. The business tends to be cyclical, and you may encounter periods when you begin to wonder if you will ever have another client. Your facilities may be very limited; as simple a resource as a good technical or business library may be unavailable to you. You will function—often after filling a role of considerable authority in industry—more by persuasion than by authority. Some people find it easy to make such a transition; others find it impossible.

Other factors important to success include being an aggressive seller, being a self starter, being goal oriented, being well organized, and being able to work on several unrelated projects simultaneously, and you will have to switch from Project A to Project F with considerable facility, all in a day's work. You will have to sell, sell, sell. The reason for emphasizing selling is that it is the function most alien to most prospective consultants, yet the one critical to success. Selling is the first step in obtaining a consulting assignment; without it, there is no consulting work.

One of the attractive features of consulting is the small investment needed to begin. Besides your brains, you need a telephone, an answering machine or service, a filing cabinet, and a desk. A computer, typewriter, and fax machine are optional. A fancy office with a fancy address is nice, but quite unnecessary. You supply all other necessities, such as the business plan, the development of contacts, and marketing yourself.

You should recognize that you will probably do everything—the planning, the marketing, the selling, and the production—and this you may well be prepared to do. But also most likely, at least in the beginning, you will do the record keeping, purchasing, transportation, traffic (mailing out samples), filing, clerical, and other chores that were previously done by other staff members.

Finding Clients

You will definitely have to market yourself. You will have to decide such things as who your target prospects are, when and how you will approach them, what your selling point will be, and what your pricing structure will be. You may wonder if any prospects are out there and how you will find them. I can assure you, that most likely they are there and that you can find them through several proven routes.

The first one is contacts. You need to develop contacts, preferably before entering the consulting field, and you need to maintain and cultivate them while building your consulting business. A second, very useful route is referrals. A satisfied client will often refer you to another division in the same company that needs a consultant and may even refer you to other companies. Visibility is very desirable, and you can attain it by writing papers, giving speeches, acting as program chairman, and giving extension courses or seminars. Such activities often result in prospects who contact you with inquiries. Referrals from organizations can also be productive. The Association of Consulting Chemists and Chemical Engineers (ACCCE) and trade associations, such as the Society of Plastics Industry or the Technical Advisory Service to Attorneys, can be sources of prospects.

ACCCE, which provides a service to its consultant members and is a resource for industry and the public, is a good example. As a member, you learn the results of its biennial survey of consulting fees and therefore get some idea of what your fee structure

PETER R. LANTOS

CERTIFICATE NO. 597

CONSULTANT TO THE
PLASTICS INDUSTRY
AND R&D MANAGEMENT

The Target Group, Inc.
1000 Harston Lane
Philadelphia, PA 19118

Telephone: (215) 233-4083

B. Ch. E. Cornell University, 1945
Ph. D. (Chemical Engineering), Cornell University, 1950

Registered Professional Engineer: New York State No. 44115.

Technical Societies: American Institute of Chemical Engineers, American Chemical Society, Society of Plastics Engineers, Society of the Plastics Industry, Commercial Development Association, Association of Research Directors, Plastics Institute of America.

Business History: President, The Target Group, Philadelphia, PA, 1980 to present; Vice President, Arco Polymers, Philadelphia, PA, 4 years; General Manager, Rhodia Plastics Div., New York, NY, 1 year; Vice President, Sun Chemical Corp., Carlstadt, NJ, 6 years; R&D Director, Celanese Plastics, Clark, NJ, 8 years; Research Supervisor, du Pont Co., Wilmington, DE, 11 years.

Engaged in Consulting Practice: Since 1980

Figure 1. Consultant's Scope Sheet.

Publications: 30+ (polymers, plastics, management science) articles; 6 patents

Scope of Activities:
Market studies and techno-economic investigations on polymers, plastics; applications, new-product positioning, product strategy.

Technical and marketing consulting: volume polymers, engineering polymers, specialty polymers, alloys/blends, advanced composites, reinforced/filled plastics, thermoplastic elastomers, films, fibers.

Expert Witness

Assistance to R&D management: organizational effectiveness, productivity, planning, creativity, managing innovation, supervisory training.

Staff: 10 Associates representing specialized skills in all aspects of polymers/plastics, management science, R&D management.

Figure 1. Continued

should be. Other major values are that it publishes a directory of members and acts as a clearinghouse for inquiries.

The directory is a complete listing of all consultant members, with a one-page scope sheet of each consultant's background and expertise (Figure 1). It also has a classifier section in which the consultants are listed by function (Figure 2), by product or material expertise (Figure 3), and by process and equipment expertise (Figure 4). The prospective client need only look under the category of product or function in which expertise is needed to find those consultants offering that expertise. This directory can be purchased and often is used by people in industry seeking a consultant.

CONSULTANT'S FUNCTIONS	7
1110 ANALYSIS & TESTING, FUSION	281 709
1115 ANALYSIS & TESTING, GAS	323 327 394 565 592 621 633 649 667 670 685 689
1120 ANALYSIS & TESTING, INFRA RED	280 489 565 591 592 621 649 664 667 670 683 685 689 705 716
1125 ANALYSIS & TESTING, INORGANIC	167 281 327 394 565 592 601 609 611 633 639 667 670 689 705 709 716
1130 ANALYSIS & TESTING, INSTRUMENTAL	281 327 394 565 591 592 621 633 639 667 670 671 683 689 692 698 704 705 709 716
1135 ANALYSIS & TESTING, MASS SPECTROMETRY	281 394 633 664 667 610 683 716
1140 ANALYSIS & TESTING, METALLURGICAL	167 293 533 565 592 668 670 698 705
1145 ANALYSIS & TESTING, MICROCHEMICAL	323 611 633 667 670 671 705
1150 ANALYSIS & TESTING, MICROSCOPIC	167 400 533 621 649 656 667 670 692 705

1155	ANALYSIS & TESTING, MYCOLOGICAL	592
1160	ANALYSIS & TESTING, NON-DESTRUCTIVE	167 565 668 670 705
1165	ANALYSIS & TESTING, NUCLEAR	281
1170	ANAL. & TESTING, NUCLEAR MAGNETIC RESONANCE	121 649 664 667 716
1175	ANALYSIS & TESTING, ORGANIC	489 534 565 584 592 611 621 633 639 649 656 667 670 683 685 689 692 705 716
1180	ANALYSIS & TESTING, PHARMACEUTICAL	280 323 592 609 611 621 651 667 670
1185	ANALYSIS & TESTING, PHARMACOLOGICAL	323 621 651 670
1190	ANALYSIS & TESTING, PHOTOMETRIC	565 656 670 671
1195	ANALYSIS & TESTING, PHYSICAL	167 293 489 564 565 591 621 649 664 670 683 692 698 704 705 709 716
1200	ANALYSIS & TESTING, PLASTICS	

Figure 2. Classifier by Consultant's Function.

MATERIALS AND PRODUCTS

PLASTICS MARKETS	496 514 541 597 622 626 656 661 673 676
PLASTICS, MEDICAL	SEE
PLASTICS PRODUCTS	293 400 489 504 541 560 564 575 591 597 622 637 649 661 664 670 676 680 683 690 692 694 696 704 705 715 719 720
PLASTICS, REINFORCED	167 293 461 504 560 564 575 581 597 626 637 649 662 664 668 670 673 676 681 690 692 705 720
PLASTICS, SANDWICH PANEL	293 461 560 564 575 649 664 676 692 720
PLASTICS, SHEETS	293 541 560 564 575 591 622 637 649 661 664 673 676 692
PLASTICS, SPECIALTY	293 400 489 495 504 560 575 581 597 622 664 673 676 680 681 683 690 692 696 704
PLASTICS, THERMOPLASTIC	167 280 293 384 400 504 541 560 565 575 581 591 597 622 626 637 649 661 662 673 676 678 680 681 683 690 692 694 696 697 704 705 706 715 719 720
PLASTICS, THERMOSETTING	167 280 293 384 400 560 564 575 581 597 622 649 657 661 662 673 676 678 680 681 683 690 692 696 704 705 706 715 719 720
PLASTICS, OTHER	293 400 489 496 541 560 575 581 591 597 622 649 673 676 681 683 690 692 706 715 720

PLASTISOLS	121 293 384 400 541 560 575 637 661 664 676 680 692 704 705 706
PLATING CHEMICALS	167 281 293 508 533 560 565 698 705 713
PLATINUM METALS	167 281 293 400 533 560 668 705 716 719
PLYWOOD ADHESIVES	280 400 541 560 656 683 692 696 716
POISONS, FORENSIC	683

Figure 3. Classifier by Consultant's Expertise in Materials/Products.

PROCESS AND EQUIPMENT

8580 SAMPLING	167 281 461 633 657 667 671 673 681 683 689 698 709
8585 SCALE REMOVAL	167 393 461 499 664 668
8590 SCREENING	499 657 664 669 689 718
8595 SEALING, CERAMIC : METAL	167
8600 SEALING, GLASS TO METAL	167 670 683
8605 SEDIMENTATION	167 499 654 664 669 689
8610 SEED DISINFECTING	SEE
8612 SENSOR TECHNOLOGY	514 698
8615 SEPARATION PROCESSES	167 477 496 499 609 613 649 653 656 657 664 667 669 683 686 688 689 692 693 697 711 717 718
8625 SEWAGE DISPOSAL & TREATMENT	167 461 499 565 601 609 657 689 693 703 711
8630 SHELL MOLDING	167 400
8635 SILK SCREEN PRINTING	SEE
8640 SINTERING	167 499
8645 SMOKE & ASH CONTROL	461 689 718
8650 SOIL STABILIZATION	280 400 690
8655 SOIL STERILIZATION	121 400 690

8660 SOLDERING	167
8665 SOLID WASTE DISPOSAL	121 167 461 499 539 601 639 656 675 676 689 692 693 703 711 718
8670 SOLUTION MINING	167 657 711
8675 SOLVENT EXTRACTION	167 280 461 633 664 683 688 689 690 697 717
8680 SOYBEAN PROCESSING	477 539 651
8685 SPRAY DRYING	SEE
8688 SPUTTERING	705
8690 SULFONATION	461 534 575 634 637 656 673 683 686 690 701 720
8693 SURFACE PREPARATION	622 664 683 705
8695 TANNING	477
8700 TEXTILES, DYEING	461 534 661 683 692

Figure 4. Classifier by Consultant's Expertise in Process and Equipment.

A more expeditious way is to call the executive director of ACCCE with a statement of the problem or need. The executive director will do a computerized search of the directory and, in just a few days give the client a list of the most qualified members. The monthly mailing to the members lists inquiries received by ACCCE, so that members who were not recommended, but feel they may be able to help, can offer their services.

Being a member of ACCCE who is listed in the directory and in the data base can be a fruitful approach to obtaining leads to consulting assignments. It should be one of several routes used. I want to emphasize that it is an approach only to leads, because you will still need to sell your services.

In summary, the successful independent consultant functions very much like a company. If you want to become a consultant, you will

- define your field by focusing and building on your strengths
- market your capabilities
- use a variety of marketing techniques, such as using contacts and getting personal referrals
- achieve visibility via publications, talks, and other activities
- join an association, such as ACCCE, that can be the source of additional referrals.

Reference

Directory of the Association of Consulting Chemists and Chemical Engineers, Association of Consulting Chemists and Chemical Engineers: New York, 1987.

Chapter 9

Opportunities for Retired Chemists

David G. Bush and John R. Thirtle

In 1982, one of us (Thirtle) published a paper in CHEMTECH magazine on opportunities for retired chemists[1]. The nominal retirement age then was 65, but most employees retired a year or two before that. Many felt the need to use their talent and experience productively in retirement, and the paper described a number of ways in which they could do so, such as volunteering or obtaining consulting contracts or part-time employment.

It was expected that legislation increasing the age for full benefits under Social Security and legislation to abolish mandatory retirement at age 65 would encourage people to work longer. Legislation extending the age to qualify for full Social Security benefits passed in 1983; the age is being increased gradually to 67 by the year 2027—not much immediate effect, if any! The legislation banning mandatory retirement at 65 was enacted in 1986; yet, the average retirement age is continuing to drop—even into the 50s. The result has been a flood of talented and experienced technical personnel into the job market. The number of consulting agencies and part-time employment firms has grown markedly.

What's Going On?

Early retirement seems to have become an attractive option to be exercised at the earliest convenient age. According to a survey in 1987 by Korn/Ferry International, 54% of senior executives stated that they wished to retire before age 65. Only 16% said they wished to work as long as possible. A bulletin from the American Association of Retired Persons (AARP) in January 1986 stated that the trend toward early retirement is expected to continue; projections show only 33% of men age 55 and older working in 1995, compared to about 31% in 1984.

Early retirement has been encouraged by employers, via sweetened benefits or incentive payments, as a means of reducing costs and increasing profits. Many of the individuals induced to retire made a career change that was beneficial; many did not. Major restructuring of industry has produced involuntary retirees who are not ready for retirement, either psychologically or financially, and are seeking new employment. The need for income has increased among those whose separation was involuntary and among those who had hoped for a career change that has not materialized in the way they had expected.

How Much Are You Worth?

One of the most difficult questions facing the retired chemist, or any other professional, is, "How much am I worth?" If you really want to earn what you are worth, you should

1. Decide what job you want.
2. Evaluate your experience and skills.
3. Estimate your worth.

You may be willing to accept less than you are worth under certain circumstances, such as:

1. The job pays a certain amount, and the rate is not negotiable.
2. You may not care about the level of pay.
3. The work may lead to other jobs that will pay more—that is, you get your foot in the door.

If you don't set a level of pay for yourself and ask for it, you may not get it. People hesitate to talk to a prospective employer about compensation and may accept much less than they are worth. For chemists, the ACS Salary Survey is one source of information about the earnings of chemists. In general, you should get payment nearly equal to the current rate. Remember, the employer gives part-time people none of the insurance and health benefits that apply to full-time employees and has little or no overhead costs related to your employment. Donald J. Berets of The Chemists Group provides some guidance in *Professional Postscripts* on setting consulting fees and other matters[2].

Temporary and Part-Time Employment

Once you have established a price on your time, what are the opportunities and options of part-time or temporary employment? Caroline Bird, in *Modern Maturity*, June–July 1988, wrote about temporary and part-time work opportunities[3]. She said

"It's hard to think of work that can't be done part-time, part-year, or on an easier schedule than most of us have to follow in our working lives. Doctors do it, lawyers do it, salespeople do it, office workers do it; and whatever work you want to do, the chances are good that you, too, can do it.

"Freedom is the advantage of part-time work. Four out of five part-timers, and even more of those who are over 55, don't want to be tied down to a regular full-time job. If they are highly skilled, they may get more per hour to make up for not getting standard benefits. Most, however, get about the same rate of pay as full-time employees but without benefits—which may account for the general impression that part-time work pays less.

"Part-time work is ideal if you would rather have extra time than extra money. You may feel this way because you already have health insurance and a pension. Or you may not want to earn enough to lose a single dollar of your hard-earned Social Security benefits."

We assume that you will want to use your skills and experience effectively. This does not mean that you will continue with your career discipline. There are many options in other areas as described by Phil Landis in *Professional Postscripts*[4]. We will describe some of them in fields such as chemistry, education, world service, and volunteering. The approach to the organizations will be fairly obvious. If you need help in job searching, we suggest you request a copy of AARP's *Working Options* guide for job-search tips[5].

Perhaps the easiest way to get part-time work is to work out a contract with your former employer, who knows of your ability, knowledge, and work habits. Many employers found that incentive separations caused them to lose more good employees than they had expected. To gain a contract, unemployed individuals may need to display some initiative. Contracts similar to those with your former employer can be worked out with other employers. Care should be taken not to violate agreements with the former employer on disclosure of confidential information. You can also apply at an employment agency for temporary or part-time work as discussed on the next page.

A good example of such a work force is the company called GLM Telesis in Rochester, N.Y. It was formed by a group of retired Eastman Kodak chemists who are skilled in the synthesis of organic compounds. They synthesize specialty organic compounds for the Kodak Research Laboratories on a contract basis, estimating the synthesis time before they start the work. Because they are former Kodak employees, they understand the need for nondisclosure of confidential information. They often work nights and weekends in laboratory space that is not used at such times.

Some Typical Consulting Firms

Retired chemists may form their own professional consulting firms or job-seeking firms. Some examples follow.

- CalSec Consultants, Inc., is composed of retired members of the California Section of the American Chemical Society and has been operating for a number of years providing fee-based and free services.

- Chelan Associates, Inc., was formed by two chemists, former congressman Mike McCormack and former ACS president Clayton Callis. The firm served as consultants in environmental compliance. (The firm is no longer in existence—publisher's note.)

- CONDUX, Inc., was formed by a group of early retirees from Du Pont to market their talents and resources.

- Career Placement Registry (CPR), a computerized job-matching service, is available to the top 8000 businesses, service organizations, and industrial companies in the United States and in 55 foreign countries who subscribe to DIALOG information retrieval service[6]. The ACS Employment Services Office provides a reduced-cost listing for ACS members.

- Intertek Services Corporation was formed in Rolling Hills, California, principally to serve the aerospace firms in that area. From that office and one in Houston, Texas, it provides temporary help to at least a score of companies in other high-technology industries. It has more than 4000 registrants. Executive offices are in Fairfax, Virginia.

- Lab Support[7] has offices in Woodland Hills, California, and in Bernardsville, New Jersey. The company is run by chemists and specializes hiring out temporary chemists for short-term staffing needs for special projects. It has a data base of more than 1000 chemists.

- Science Temps[7] has more than 3000 chemists, biochemists, pharmacists, and engineers as registrants. It provides consulting a well as extended assignments. The assignments are usually for three to four months, and the work week averages 37½ hours. Companies find this an effective way to get expertise. The company is planning to open an office in Philadelphia.

- Second Careers, Los Angeles, California, maintains a skills bank of applicants. It screens and counsels applicants and also, acting as an employer, handles payroll services for clients, thus allowing companies to rehire retired employees on a temporary basis without affecting their pension status. Many communities have similar organizations.

Telephone books in major cities or your local chamber of commerce can probably provide contacts with such organizations in your area. If you can't find a local service that handles your skill, send a self-addressed envelope to the National Association of Temporary Services[8]. It will send you a list of employment services in your area.

What Corporations Are Doing

Some corporations are providing innovative work options for continuing employment of older workers and for rehired employees. An article in *Aging* documents 369 innovative company practices uncovered in a study by the Institute of Gerontology at the University of Michigan with support from the Administration on Aging[9]. The article was designed to address the need for information exchange among companies on a variety of existing personnel practices for older workers. The study showed that altruistic motives may have stimulated practices helpful to older workers, but that the practices were continued because they were advantageous to the employer. Single copies of U.S. Senate report No. 99–10, which covers this study, are available[10].

Another project that will interest employers of older workers is the Worker Equity Initiative of AARP. It has three main thrusts: to develop and disseminate new information about the abilities, needs, and rights of older workers; to educate employers, employees, and the public about those abilities, needs, and rights; and to implement advocacy programs aimed at changing workplace patterns that impede the legitimate employment interests of older persons. The initiative is administered by the Worker Equity Department of AARP. The department publishes a bimonthly newsletter, *Working Age*, about employment issues affecting middle-aged and older workers. The newsletter is written for personnel, benefits, and human resources professionals and training directors[11].

The Trends

In the *Wall Street Journal* for June 6, 1988, Amanda Bennett wrote of the current situation regarding free-lance workers[12]. Part-time and temporary workers make up at least 25% of the work force, and this so-called contingent employment is growing fast. Following the 1982 recession, the use of temporary help grew about 25% per year. These conditions are expected to continue. They allow companies to cut costs quickly when business slackens. Labor groups are apprehensive, however. They cite instances of full-time workers being fired and immediately being rehired as part-timers without benefits and at lower salaries.

Concerns of Employers, Labor Officials, and Politicians

Corporate and labor officials, politicians, and analysts have various concerns about employers' treatment of part-time and temporary employees. For example,

- Should companies be compelled to give free-lance workers the same health-care benefits their regular workers get?
- Should legislation specify that part-time workers be paid at a pro-rated full-time rate?
- Should companies be required to design pensions so that mobile workers can take them from one job to another?
- Should restrictions be placed on the ability of companies to fire full-time workers and replace them with a staff of part-timers?

In 1988, Rep. Pat Schroeder (D.–Colorado) introduced a bill in Congress that would provide for mandatory pro-rata pension and medical benefits for any employee working fewer than 30 hours per week.

Volunteering

If your retirement income is sufficient for you and your family, you may find volunteering to be satisfying. Most communities need volunteer aid and generally advertise for such help from individuals. Some communities maintain a centralized volunteer data bank that helps to match the skills of volunteers with the needs of organizations. In many communities, art galleries and museums provide rewarding uses of your time. If you have technical skills, you may find that your local public broadcasting station will provide activity that is challenging and rewarding. In many cities, a section of the telephone book is devoted to government and local agencies that can use your talents.

The ACS has put together a skills data bank of volunteers that is accessible to national agencies that need such help. Also, AARP has a Volunteer Talent Bank of more than 300,000 volunteers who provide a variety of services. This volunteer force works through local chapters on a variety of community projects.

Volunteering with Financial Support

Some programs for volunteers offer financial support. Here are a few of them:

- The Council for International Exchange of Scholars grants Fulbright–Hays awards for university lecturing and research abroad.
- The Forty-Plus Club finds jobs for senior executives and professionals. The club is headquartered in Washington, D.C., and has offices in major cities.
- The International Executive Service Corps (IESC) recruits retired, highly skilled U.S. executives and technical advisors to share their know-how and expertise with business and technical people in developing nations. It has a skills bank of about 8500 men and women and has successfully completed about 9000 projects in 74 different nations since 1965. Each project generally lasts between two and three months. Volunteers are encouraged to travel overseas with their spouses. IESC pays the travel expense and a comfortable per diem allowance, but no salary.
- International Voluntary Services, Inc. (IVS), provides technical assistance in rural, nonindustrialized countries.
- The Peace Corps recruits volunteers for overseas service (two years) in a variety of projects including teaching of math, biology, chemistry, and general science.
- RSVP (Retired Senior Volunteer Programs) assists public and nonprofit institutions. Volunteers are reimbursed for travel and meal expenses.
- SCORE (Service Corps of Retired Executives) provides managerial or even technical services to businesses.
- VISTA (Volunteers in Service to America) recruits volunteers for community projects for one to two years in the United States. Volunteers receive a small salary to cover living expenses.

- VITA (Volunteers in International Technical Assistance) provides solutions to technical questions related to third-world development.

Some Volunteering Opportunities Requiring Technical Input

Volunteering can lead to consulting work, if you volunteer your time in fields close to your technical expertise. In this way, you can learn of local technical needs and use that knowledge as a form of networking. One of us (Bush) has found this to be true in Rochester, New York, where he continues to live after retiring from Eastman Kodak Company. He now offers a consulting service in occupational health and environmental chemistry.

Upon retirement, Bush did volunteer work for the following organizations:

1. The Center for Environmental Information (CEI), a nongovernmental center for environmental information. The center does not take positions on the various issues. Bush serves on its solid waste issues committee, which is planning public education programs on reduction, reuse, and recycling of waste.

2. The Chemical Hazards Information Team (CHIT), founded by the Rochester Section of the ACS and the Monroe County Office of Emergency Preparedness to train for and assist in the event of a chemical release.

3. Life Line, a 24-hour telephone poison control center as well as a source for short-term supportive counseling and information about human services. Bush is one of Life Line's consultants on problems involving hazardous materials.

4. Service Corps of Retired Executives (SCORE), a group of retired volunteers dedicated to advising small businesses. Bush recently advised a client who was purchasing a dry-cleaning establishment.

Some Volunteer Teaching Opportunities

Rochester offers many opportunities for volunteers as teachers,

which can also have networking value. Bush taught classes for the following programs:
1. Chemical Education for Public Understanding Seminars (CEPUS), sponsored by the local community college, which, for example, gave firemen a better understanding of chemical fires and garden store employees a better understanding of the use of pesticides.
2. Chemical hazards in the environment, a topic for which teaching seminars and classes are needed at the college level.
3. Science and Technology Entry Program (STEP), commencing in the seventh grade, for economically disadvantaged students who have an interest in science. The program has activities every Wednesday afternoon and Saturday morning, as well as during February and April recesses.

Some Federal and State Regulations

Bush's present consulting practice is concerned, for the most part, with helping clients come into compliance with federal and state regulations. Below is a list of those regulations where his consulting work has been useful, particularly for smaller businesses that do not have full-time technical people on their staffs.

- HCS (Hazard Communication Standard [of the Occupational Safety and Health Administration]) is for employers whose employees use hazardous materials that require labeling, training, and Material Safety Data Sheets. HCS is commonly known as the Employee Right-To-Know Law.
- SARA (Superfund Amendments/Reauthorization Act of 1986, Title III) establishes a framework for communicating hazards associated with chemical use in the workplace to the community in general. SARA is commonly known as the Community Right-To-Know Law.
- RCRA (Resource Conservation and Recovery Act) is a program designed to protect human health and the environment from improper hazardous waste management practices. In 1986, RCRA was extended to small waste generators, defined as

Table I. Consulting Opportunities (Past Two Years) for David G. Bush Associates

Client	Hazard Communication Standard— Employee RTK[a]	Superfund Amendment/SARA— Title III Community RTK[a]	Resource Conservation Recovery Act	Industrial Hygiene (PELs)	Environmental Audit	Other
Businesses						
Hardwood furniture	X	X	X	X		
Machine gauges	X	X	X	X	X	
Power supplies	X	X	X			
Circuit boards	X	X	X			
Storage racks	X	X	X			
Schools						
Suburban 1	X		X	X		
Suburban 2	X	X				
Suburban 3	X	X				
Suburban 4			X			
BOCES[b]	X					
Urban	X		X	X		X
Consulting						
Local engineering					X	
National environmental					X	X

NOTE: The two consulting firms are Larson Engineers (in Rochester, N.Y.) and Chelan Associates (in Washington, D.C.).
[a]RTK means right to know.
[b]Board of Cooperative Educational Services.

companies that generate between 100 and 1000 kilograms of waste per month.

- PELs are permissible exposure limits for toxic materials that should not be exceeded in workplace air. PELs are set by the Occupational Safety and Health Administration.
- Environmental assessments, or property transfer audits, should be completed before commercial property is purchased. Of significant concern to landowners is their liability for the cost of the cleanup of hazardous waste under the Superfund (CERCLA) law.

Table I shows some specifics of Bush's consulting practice.

Summary of Consulting Income

Table II shows Bush's consulting income over the past five years as a percentage of his income during full-time employment. His earnings are only a fraction of his former salary in industry, but he wants it that way. In retirement he can control his own life and work only part time. Note also the negative income during the first year, when startup costs exceeded consulting income.

Table II. Summary of Consulting Income for David G. Bush Associates

Year	Gross	Net
1984 (9 months)	0	−5
1985	8	4
1986	19	10
1987	21	8
1988	28	13

NOTE: As a percent of salary earned as an analytical chemist in industry.

References

1. Thirtle, John R. *CHEMTECH* **1982**, 652–655. Reprinted in *Financial Planning, A Guide for ACS Members*, by Mark R. Greene, sponsored by and underwritten by the Board of Trustees, Group Insurance Plans for ACS Members, 1988.

2. Berets, Donald J. In *Professional Postscripts* **1988**, Spring. Published by the ACS Office of Professional Services, Washington, DC 20036.

3. Bird, Caroline *Modern Maturity* **1988**, June–July, 38.

4. Landis, Phil In **1988**, Spring.

5. Obtain AARP's *Working Options* guide for additional job-search tips. Request a single copy from Working Options, AARP Fulfillment, P.O. Box 2400, Long Beach, CA 90801.

6. *Professional Relations Bulletin* **1987**, May (39). Published by the Division of Professional Relations, ACS, 1155 16th Street, NW, Washington, DC 20036. Forms may be obtained from the ACS Employment Services Office at the same address.

7. *Chemical Week* **1986**, Aug. 20, p 24.

8. For a list of employment services in your area, send a self-addressed envelope to the National Association of Temporary Services, 119 South Saint Asaph Street, Alexandria, VA 22314 (or call 703-549-6287).

9. Root, L. S. "Corporate Programs for Older Workers" *Aging* **1985**, No. 351.

10. "Personnel Practices for an Aging Work Force: Private Sector Examples", Report No. 98–10, by L. S. Root and L. Zarrugh, is available from the Special Committee on Aging, U.S. Senate, SD–G33, Washington, DC 20510.

11. The newsletter may be obtained without charge by sending a request on company or organization letterhead to Working Age, AARP Worker Equity Department, 1909 K Street, Washington, DC 20049.

12. *Wall Street Journal* **1988,** June 6, p 24.

Some Typical Consulting Firms

- CalSec Consultants, Inc., California Section, ACS, 511 Wells Fargo Building, 2140 Shattuck Avenue, Berkeley, CA 94704. Phone 415-848-0513.
- CONDUX, Inc., 228 Unami Trail Newark, DE 19711. Phone 302-738-9200.
- The Chemists Group, Inc., Box 3365, Ridgeway Station, Stamford, CT 06805. Phone 203-322-9210.
- GLM Telesis, 38 Thorntree Circle, Penfield, NY 14526. Phone 716-377-3579.
- Inter-American Institute of Science, Inc., 2000 South Washington Avenue (Suite 2), Titusville, FL 32780–4739. Phone 305-269-3221.

Agencies for Part-Time Professional Employment

- ACS Employment Services Office, 1155 16th Street, NW, Washington, DC 20036. Phone 1-800-227-5558. (In Washington, DC, area, phone 872-4528.)
- Association of Part-Time Professionals, Flow General Building, 7655 Old Springhouse Road, McLean, VA 22102. Phone 703-734-7975.
- Career Placement Registry (CPR), Wilmington, DE 19880. ACS members can apply via the ACS Employment Services Office.
- Intertek Services Corporation, 9900 Main Street, Fairfax, VA 22031. Phone 703-591-1320.
- Lab Support, Inc., 20301 Ventura Boulevard, Suite 305, Woodland Hills, CA 91364-2447. Phone 818-716-8990.

- GLM Telesis was formed by three retired chemists from Eastman Kodak as described in the previous section.
- The Chemists Group, Inc., was formed by a group of retirees from American Cyanamid. Its mission is to find temporary or consulting employment for experienced chemists and chemical engineers.
- The Inter-American Institute of Science involves permanently retired people. The organization is not yet operational but plans to offer analytical services and applied research capabilities supplied by retired men and women to corporations, government agencies, educational institutions, and private organizations on a fee basis. The fees will be shared with the employees.

These are examples of the types of organizations that are evolving. It should be recognized that some are composed of former employees of a single corporation and will probably confine their services to their regional communities but will be willing to receive inquiries from individuals and groups interested in establishing similar organizations.

Agencies for Part-Time Professional Employment

Employment agencies specializing in part-time employment of chemists have been growing in number and in size. Examples follow:

- Association of Part-Time Professionals (APTP) is a national organization with members throughout the United States in foreign countries. Its aims are to promote opportunities for qualified men and women in part-time professional work and to bring together employers who require professional skills with professionals who prefer flexible employment. APTP advocates prorated salaries and employee benefits for part-timers.
- ACS Employment Services for ACS members provides listings for temporary and part-time, as well as full-time, work. To use the service, contact Employment Services (see references). The service may also be contacted for listing in the Career Placement Registry (see the next item).

- Science Temps, 108 North Union Avenue, Cranford, NJ 07016. Phone 201-272-1997.
- Second Careers Program. Situated in Los Angeles, CA. Maintains a skills bank; typical of programs in the major cities. Consult your local aging agencies and the United Way.

Volunteering with Financial Support

- Council for International Exchange of Scholars, Suite 300, 11 Dupont Circle, NW, Washington, DC 20036. Phone 202-939-5400.
- Forty-Plus Club. Corporate office in Washington, DC. Offices in major cities.
- International Executive Service Corps (IESC), 8 Stamford Forum, P.O. Box 10005, Stamford, CT 06904-2005. Phone 203-967-6006.
- International Volunteer Services, Inc., 1424 16th Street, NW, Washington, DC 20036. Phone 202-387-5533.
- Peace Corps, 806 Connecticut Avenue, NW, Washington, DC 20525. Phone 202-254-6886.
- Retired Senior Volunteer Program (RSVP), ACTION, 806 Connecticut Avenue, NW, Washington, DC 20525. Phone 202-254-7346.
- Service Corps of Retired Executives (SCORE), 1128 20th Street, NW, Suite 410, Washington, DC 20036. Phone 202-653-6278.
- Volunteers in International Technical Assistance (VITA), 1515 North Lynn Street, Arlington, VA 22207. Phone 703-276-1800.
- Volunteers in Service to America (VISTA), ACTION, 806 Connecticut Avenue, NW, Washington, DC 20525. Phone 202-254-7346.

Chapter 10

Using Consultants To Interpret Regulatory Initiatives

Mike McCormack

In 1985, Clayton Callis suggested that we form a company to provide consulting services in environmental compliance and human health and safety. Callis, who had retired from Monsanto Fiber and Intermediates Company, was acutely aware of the need for such consulting services, especially among small companies or large companies with many subcontractor suppliers of chemical intermediates. This need was created by the extraordinary increase in regulations promulgated because of new or newly amended federal and state laws, such as the Occupational Safety and Health Act and the spectrum of environmental protection laws enacted by the U.S. Congress. Thus, Chelan Associates, Inc., was created.[1]

Organizational Steps

One advantage we had in organizing the firm was that I was already incorporated (as McCormack Associates) and was providing consulting services in science, energy, and government. I had a functioning office with a secretary and manager and all the essential

---

[1]Note: After this chapter was prepared, Chelan Associates, Inc., was dissolved by McCormack and Callis.

equipment for providing consulting services. We were paying taxes, social security payments, workman's compensation, and so forth. In short, McCormack Associates was already doing what a small corporation must do to function and comply with state and federal laws and regulations. Instead of duplicating all of this, Chelan simply purchased all of its office and administrative services from McCormack Associates. The benefit is that Chelan Associates has no employees, in the legal sense, and owns no equipment. Callis and I are paid as consultants.

We invited Marvin Glaser, recently retired manager of the environmental affairs program at Exxon Research and Engineering Company, in New Jersey, to join us as a vice-president. We also invited Doug Costle, former Administrator of the Environmental Protection Agency (EPA), and David Berz, environmental attorney in the legal firm of Weil, Gotshal, and Manges, to become members of our board of directors. We kept the corporation private, with stock ownerships limited to board members.

We assembled a team of supporting experts in the various scientific disciplines related to environmental compliance and human health and safety. In 1985, a large reservoir of experienced experts was available. They included professionals who had taken early retirement from executive positions in industry, members of university faculties, and individual entrepreneurs already in the field.

We began building our team by inviting acquaintances we had made in industry and through activities in professional societies. In this manner, we obtained several outstanding associates. We also searched the American Chemical Society file of retired chemists, selected about 24 people from about 1200 in the file, and invited them to participate. Several of them accepted and recommended additional candidates. Some declined, but suggested alternatives. Next, we examined the list of potential associates on the basis of experience and discipline and then sought qualified individuals in disciplines not already covered. In time, we assembled a team of about 24 associates with skills in all required disciplines to respond to our potential clients' needs.

During the next three years, we added several new team members and several withdrew. In addition to associates with recent experience in conducting audits for compliance with environmental and occupational safety and health regulations, the associates bring to the organization, among other skills, expertise in the following areas:

- industrial waste management
- nuclear waste management
- analytical chemistry
- chemical process and information management
- pollution control technology
- human health and safety, industrial hygiene
- food and drug chemistry
- radiochemistry
- science and law
- toxicology
- indoor air pollution
- unusually hazardous organics management
- environmental compliance management
- knowledge of government procedures
- technical education
- laboratory safety
- research engineering management
- air quality control and particle and aerosol emissions

The Chelan Associates are, of course, individual entrepreneurs—on call when we agree with them on a job and remuneration for it.

Market Characteristics

We soon learned that the phenomenon that helped make experienced experts available to join us (i.e., reduction in force in the chemical industry) also produced competitors for us. We also learned that major corporations that we might have normally expected to be customers instead developed competitive environmental organizations from among their personnel who might otherwise have been terminated. These developments confronted us with more qualified competition then we expected and, at the same time, reduced the market that we anticipated among large corporations.

We found that most small corporations—who needed our assistance the most—concluded that they could not afford our services. In general, they decided simply to gamble that EPA would not catch them in violations of regulations. Therefore, the most obvious market segment available to us was medium-size corporations—those big enough to have management sufficiently sophisticated to recognize the need, but too small to afford, in-house, the competence we could provide.

Marketing Methods

The medium-size companies that seemed our most likely market can be reached only with aggressive marketing and selling—a skill generally outside the competence of our company. The point is a major one for those considering creating a small consulting company to provide services in a scientific discipline. Unless clients are available through previous association or personal relationship, a marketing and selling campaign is essential if one is to obtain customers. Such a campaign requires either a significant initial investment or a modest guaranteed income with additional incentives for the person or organization doing the marketing.

We prepared a marketing brochure to tell our story. Some quotes from it are

> Today, thousands of small and medium-size U.S. companies are confronted by a tidal wave of new laws and regulations related to the production, use, handling, or disposal of a long list of chemicals—many of which are newly classified as hazardous. The effect of these laws and regulations is to impose demanding new responsibilities on businesses not prepared or qualified to handle them. Failure to comply with these requirements may result in severe penalties, costly litigation, or both.

> Chelan Associates brings you a team of professionals with extensive industrial experience in environmental management and compliance. We offer the benefits of this mature judgment and sympathetic understanding of your problems—to provide you with a critically important advantage. Our expertise in environmental compliance—and in the broad spectrum of scientific disciplines required to master all aspects of regulations related to chemicals—can be of immediate and decisive assistance to you in understanding and complying effectively with environment laws and regulations.

Expanding on this:

Wherever your facilities are located, whatever your chemical operations may be, Chelan Associates can act as your legal and technical research arm in environmental compliance: quickly, professionally, and economically. We can help you understand what the law requires of you, and how best to achieve your goals of improved performance with regard to environmental compliance and human health and safety.

The brochure also stated that Chelan Associates can:

1. scope your unique requirements with respect to compliance with environmental laws and regulations
2. conduct systematic, comprehensive environmental reviews of your facilities and operations
3. provide guidance for you to develop your own low-cost environmental self-audit program to meet your needs for your facilities and operations
4. evaluate your plans for remedial action to eliminate deficiencies or weaknesses in environmental law compliance.
5. make recommendations for your management plan—with high standards of safety and environmental compliance
6. assist in the preparation of manuals and special documents as required by law and regulations. For example
    a. spill control manuals
    b. training manuals
    c. hazard communications plans
7. provide guidance for compliance with anticipated federal and state laws and regulations, or amendments to existing ones

8. train your environmental staff
9. assist in the education of employees with respect to chemical hazards (as required by law)
10. provide general consultation on technical and legal aspects of your environmental control and compliance requirements; and provide expert, professional technical testimony for legal proceedings and litigation

Insurance Problems

One frustration has been that Chelan Associates has been unable, to date, to purchase directors and officers insurance or professional liability insurance. We have stipulated in writing that we would not do any physical work, touch anything, supervise any work by anyone outside our company, recommend any engineering companies, and, in short, do anything but advise clients of the results of our audits. Nevertheless, the mention of the words environmental regulations or chemicals or health and safety seems to preclude any opportunity for us to obtain insurance. We are continuing to search.

At the same time, we are seeking to protect ourselves by including in our contracts with our clients indemnification provisions against third-party lawsuits. In this effort, as might be expected, we have been only partially successful, and we have been forced to gamble that we will not be involved in any such litigation. Of course, all of our work and our reports are confidential and protected as attorney–client privileged documents.

Examples of Work

Our clients represent an unusually broad spectrum in size and activity. Some examples are discussed next.

A Small Manufacturer. We found a very small manufacturing operation involving etching and potting, using a variety of acids and organic chemicals. The company was in substantial compliance with state and federal regulations, but we learned that it had stored many drums of liquid waste without any clear knowledge of the contents of the drums. Management was surprised at the intensity of our concern, but accepted our recommendation to have the drums legally removed.

A Large Office Building. When checking for air pollutants in an unusually clean office building, we discovered that a pristine room (of which management was very proud) contained near-hazardous concentrations of carbon dioxide. No one could understand why until we found that the new ventilation system did not operate unless the heat or air conditioner was on. The carbon dioxide build-up was caused by operations in the room.

A Large Petroleum Company. We were asked by a large petroleum company to confirm environmental audits on several facilities. This was a major and difficult job requiring teams of several specialists at each of two sites. We reported that the environmental audits prepared in-house were not completely realistic.

A New Food Product. We conducted environmental audits of the facilities in which a large company planned to manufacture a new food product. We assisted in preparing a license application to the U.S. Food and Drug Administration.

A Real Estate Transaction. We completed a simultaneous audit of nine small newspaper printing plants, all of which were involved in a real estate transaction.

Supporting Work. We also assisted clients by providing accurate, understandable, relevant information for use in litigation or congressional testimony.

In our work on these and other projects, we used as many as five specialists on one job. Our clients benefit from their interaction on site, and we enter and leave a plant quickly. We give management an oral report of the key findings of our audit before we leave a site. The oral report is followed by a confidential written report, usually submitted within a week.

The response we have received from our clients indicates that they are pleased with our work. We think we are doing a good job at providing the expertise, experience, judgment, and sympathetic understanding that our balanced team of associates offers.

Organizational and Operational Essentials

My advice to others contemplating entering upon a similar operation is to

1. incorporate and explore the advantages and disadvantages of S-Type corporations
2. include an attorney skilled in corporate and tax law on your board of directors and as a stockholder
3. obtain directors and officers insurance and professional liability insurance if possible
4. establish or obtain a marketing competence. recognize that someone must almost certainly spend much time selling your company's services.
5. maintain the highest standards of toughness, confidentiality, and integrity in helping your clients comply with all laws and regulations involving environmental compliance and human health and safety.

Chapter 11

Major Chemical Company Retirees as Consultants and Market Developers

Robert W. Belfit, Jr.

The past several years have seen much restructuring and shrinking of staffs in all industry. It appeared to some of the people who later founded Omni Tech that, although many jobs were being eliminated, the functional needs still existed. We also realized that experienced people hired as consultants would pose only a short-term expense. Several of the founders attended a seminar entitled "Starting Your Own Business" provided by the Dow Chemical Company for those of us who were contemplating retirement. During this seminar and some of the work sessions therein, three of us looked at each other and said, "Why don't we start our own company to market the skills of a variety of Dow retirees?" It occurred to us that most people do not market themselves well, even though they possess expertise that would be very much in demand. The facilitator of this seminar encouraged us strongly, saying that this was a very new concept to him.

Formation of the Company

With these thoughts in mind, six executives, all former chemists or chemical engineers, retired from the Dow Chemical Company and formed Omni Tech International, Ltd.

- Richard K. Avitabile, vice-president of marketing, has broad marketing, sales, and business experience. Among recent responsibilities, he managed a large portion of the inorganic business segment of the Dow Chemical Company.
- Robert W. Belfit, Jr., president and chairman, spent his Dow Chemical career in product, process, and analytical research with the last seven years as manager of quality standards for the corporation.
- James H. Hanes, executive vice-president and secretary, is responsible for our legal matters. He was formerly plant manager for Dow's plant in Rocky Flats, Colorado. Before retiring, he was a vice-president of Dow Chemical USA. His responsibilities included employee relations, general counsel, and insurance.
- Richard L. Heiny, vice-chairman, was business manager of a large segment of the organic chemicals department for the Dow Chemical Company. In addition, he has extensive research and development experience.
- Robert L. Hotchkiss joined Omni Tech as executive vice-president with responsibility for technology. His entire Dow career was spent in research and development covering almost every aspect of inorganic, polymer, and organic chemistry.
- The final member of the six founders is Winfred C. Zacharias, executive vice-president for finance and administration. He started as a chemical engineer but soon moved into treasury, insurance, and finance. He finished his career at Dow as assistant treasurer for the U.S. area.

Within 12 months of our founding, two other people joined us as officers of the company:

- Larry L. Rice assumed the role of director of sales. He was director of marketing for the inorganic chemicals department of Dow Chemical USA before retiring.

- Marvin E. Winquist became Omni Tech's regional director for the Gulf Coast, where a number of our associates are located and a growing number of clients is developing. He has extensive experience in research, development, and business management in Europe and the United States.

During the first six months of 1986, we conducted seminars outlining Omni Tech in California, Ohio, Texas, Louisiana, and Michigan and listened to groups of 6 to 12 people telling about their experiences and what they wanted to do with their knowledge. By December 1986, we had exceeded our early goal—to have 50 experts under contract to perform consulting services for our clients. Today, we have more than 140 such associates. This list could well be much larger. In fact, 150 resumes are in our list of potential associates. This list is also available to match up with client needs as necessary.

By December 1986, we also had a group of marketing associates across the United States, and we had indications from several people in Europe and Canada that they would like to become associated with us. Most of our marketing associates are former Dow sales people. They include former Dow corporate account managers and general sales managers as well as a small group of non-Dow retirees.

The criteria for selection of our associates, whether they be marketing or technical associates, are similar. First, they must possess technical excellence and competence in their fields of expertise, whether the field is computers, polymers, marketing, employee relations, or other skills. Second, they must possess the ability to adapt their technology and experience to our clients' needs.

In our very earliest discussions leading to the formation of Omni Tech, we developed a mission statement: "To assist clients in conducting their affairs more successfully through the use of Omni Tech's resources." This is a sincere, broad statement, but that's what mission statements are supposed to be. Nevertheless, it is necessary under such a statement to develop some corporate objectives as shown in the following list:

1. to provide quality services to our clients at values that the clients perceive to be desirable
2. to provide rewarding opportunities for Omni Tech associates

3. to participate in other ventures that offer opportunities for Omni Tech and our associates
4. to market selected products that are complementary to our services
5. to accomplish these objectives in a manner that reflects positively on our communities, associates, employees, and ourselves

Work Performed

Following are some examples of Omni Tech's work that reflect our first corporate objective.

Two of our analytical specialists analyzed and evaluated an instrument for measuring the breath of a patient during and after surgery. This instrument was reputed to be capable of analyzing for three different anesthetics, plus oxygen, nitrogen, and carbon dioxide simultaneously within one-tenth of a second. We feel comfortable that the venture capitalists' firm was satisfied with our findings.

A medium-sized chemical company was considering the acquisition of a plant in Texas. The company was interested in examining this plant from an environmental standpoint to see if, by any chance, it was buying a big black hole that would consume much money. One of our associates audited the plant thoroughly and wrote a report that pleased this company so much that it invited him to come back and do a complete audit of all aspects of the potential acquisition. An appropriate decision was made, based largely on our associate's report. Our associate, by the way, was formerly a general manager of one of the major manufacturing divisions for the Dow Chemical Company.

We have done marketing research to determine the potential market for instruments capable of measuring the concentrations of several sensitive materials in the environment. Analyses were compiled of competitive equipment in the marketplace, and market opportunities were defined for the client company and its line of instruments. The environmental regulations were also assessed relative to the use of these instruments.

We have assisted several companies with employee relations problems. In most cases, these requests seem to come at critical times, requiring night and day and, sometimes, weekend efforts to understand the situation and to develop an appropriate program for the management and employees involved. Generally, this work includes development of appropriate safety programs, communications programs, and employee manuals. In every one of these cases, a win—win situation evolved for all people concerned.

A very rewarding project occurred in Michigan. A third-tier supplier to the automobile industry flunked a quality audit by a second-tier supplier. This failure resulted from lack of an appropriate quality assurance program. We were called in to help by the third-tier supplier. We visited the plant one afternoon to analyze the situation. As we were preparing to leave, the owner and president asked if one of our people could help him on another problem. He showed us several letters written by the city engineer, the city manager, and the city council. They pointed out that the wastewater from his plant was going into the city waste treatment plant and causing its sludge to be so loaded with metal cations that it had to be considered hazardous material.

I immediately telephoned one of our environmental associates, and the next day we returned to the plant. We examined the plant's waste disposal system and identified the problems. We attended the council meeting with the city manager and engineer and were able to convince the appropriate city officials that the situation could be corrected in the very immediate future.

Well, you can anticipate the ending of the story. We did the waste disposal job; we provided training for our client's employees in handling hazardous materials; we completed the quality assurance job; and the second-tier supplier is now bringing his business back to our client's plant, which otherwise would probably have been shut down. As a measure of his satisfaction with our performance, the owner of this company has referred other business to us on several occasions.

Another story I titled "Responsiveness." One Thursday afternoon, I received a phone call from a person asking specifically for me. It was someone whom I had worked with 15 or 20 years ago. This person had purchased an older, East Coast manufacturing plant whose product was in short supply and great demand. The plant was running at 50% or less of rated capacity. Product quality apparently was satisfactory, but management knew it had to invest some capital to get the plant up to date. Frankly, they did not want to do this until a knowledgeable process engineer had gone through the plant and made appropriate recommendations. By Friday afternoon we had a proposal in the mail to them. On Tuesday the proposal was approved, and by Wednesday one of our most capable process engineering associates was on a plane. He spent two days examining that plant and wrote a report with a number of recommendations that pleased our client. We are looking forward to another two to three weeks of work assisting that client in implementing the recommendations.

A final success story involved a small specialty chemical plant in the East (a very successful, busy, absolutely first-class operation). Management asked us to evaluate the plant's wastewater permit, to develop a spill prevention program, and, finally, to help the owners find a new plant location. We identified and examined 26 potential sites and narrowed the choice to three for which we did economic evaluations and cash flow studies. The client accepted our recommended first choice; we are now developing the appropriate environmental permits and assisting the owners in applying for financing assistance needed to move the plant to the chosen site in a neighboring state. We have developed a complete architectural projection of the new plant on the new site, including the landscaping, the sewer and waste disposal system, and the storm sewers. In the near future we expect that all the approvals will have been secured, and we will begin the process design for this plant, follow the construction, and later, assist during start-up.

Somewhat to our surprise, 25% or more of our activities have been in marketing research. We have what we think is a unique and valuable approach to marketing research projects. We put together a team made up of a technical expert along with a professional marketing researcher. For instance, we have teamed an analytical research scientist with a professional marketing researcher on an analytical marketing research project. We believe this approach gives us an advantage over more traditional means of conducting market research.

Several activities reflect our second corporate objective: "To provide rewarding opportunities for Omni Tech associates." Our consulting associates are independent contractors, most of whom want to work part time. Eighty-five percent are ex-Dow people, and they are geographically spread, with 84 in Michigan, 26 in Texas, and 28 scattered across the rest of the United States.

Our standard operating practice is that Omni Tech identifies and defines the project and negotiates the business agreement with the client. We obtain commitments from our consulting associates, and we provide the support and guidance needed to carry out the project. The associate usually helps to present the results to the client. Omni Tech bills the client and pays the consultant.

Our associates are as important to us as are the paying clients. When we look at the skills available through our resource group, we find that they fall naturally into the 10 specialty groups shown in the following list.

1. business and marketing
2. computer technology
3. employee relations
4. insurance and financial
5. product knowledge
6. production knowledge
7. quality management
8. regulatory and environmental
9. safety, security, and loss prevention
10. scientific and technical skills

Each of these specialty groups has an Omni Tech principal or officer as a sponsor to provide general leadership and ensure consistency of operation.

Our generic brochure covers the general background of Omni Tech associates and the 10 categories of expertise described in the previous list. Specific brochures have been developed for each of these 10 categories. Thus far, some 70 of our associates have completed one or more projects in about three dozen states.

Other Ventures

Our third objective is "To participate in other ventures which offer opportunities for Omni Tech and our associates." In this vein, the following news hit many publications in March of 1988.

Omni Tech Launches Product Line

Omni Tech International Ltd. has been granted an exclusive license by the Dow Chemical Company to develop and market ion exchange polymers based on Dow's ethylene–acrylic acid copolymer resin. The ion exchange polymer scavenges metal ions from aqueous solutions.

This arrangement is a new approach to developing and marketing a product that would otherwise have been shelved by Dow as another interesting discovery that did not quite qualify for the resources necessary to make it through the exacting requirements of a modern giant corporation. In this case, Dow had determined that the market was too small and too diverse to fit its marketing organization. When we realized that this product was being mothballed, we decided to try to license it. A number of our associates had been in the ion exchange business as researchers and technical development people. Two of Omni Tech's founders had spent parts of their careers in the ion exchange business, and other associates had relevant expertise in environmental matters and manufacturing. A business that appeared to be too small for Dow seemed just the right challenge for our group of entrepreneurs.

Our people were also excited by the opportunity personally to share in the risks and rewards of starting a new business. Besides obtaining the agreement with Dow, Omni Tech has provided the financial basis, facilities, and management; the associates are investing their time and talents in anticipation of being well rewarded when the business becomes a commercial success. The arrangement with Dow gives Omni Tech the exclusive worldwide license under Dow's patent rights and trade secrets. More than 15 people are now actively involved in developing markets for this product. As indicated, their rewards are obvious: a "piece of the action" based on their individual participation.

Complementary Products

Our fourth corporate objective—"To market selected products which are complementary to our services"—was established to ensure direction for another of our activities.

One venture under this objective is the marketing of Hazox Corporation's software for producers, consumers, formulators, and distributors of chemicals considered hazardous. This description covers almost everybody, as all chemists know. The Hazox Corporation's software is designed to assist users in managing their responsibilities under the Occupational Safety and Health Act (OSHA), the Resource Conservation and Recovery Act (RCRA), and the Superfund Amendments and Reauthorization Act (SARA).

Toxic Alert, Hazox's basic software system, uses a master menu to access 10 functional modules. After assessing several systems, we believe the Hazox system to be the best, as it is user-friendly, is focused on regulatory needs, and is updated as needed. We were impressed by the company's staff.

Quality in Performance

Our fifth corporate objective, which almost speaks for itself, is "To accomplish these objectives in a manner which reflects positively on our communities, associates, employees, and ourselves." We have offices in Midland, Michigan, and Lake Jackson, Texas. We are active in our respective chambers of commerce. We employ two full-time secretaries and an accountant co-op. Omni Tech has purchased four computers and other office equipment. We lease 3000 square feet of office space in Midland and about 1000 square feet in Lake Jackson.

The extent of our activities is evident in the following partial list of types of projects anticipated for the near future.

1. market research
2. scientific and technical
3. environmental audits
4. employee relations
5. environmental and regulatory
6. agricultural product development
7. computer systems design and programming
8. technology assessment

9. cogeneration technology
10. mixing technology
11. aircraft thermal insulation
12. analytical service
13. market analysis
14. laboratory accreditation
15. expert witness
16. aquatic toxicology
17. production of plastic materials
18. motor vehicle painting process
19. waste disposal
20. quality management programs
21. waste-to-energy conversion
22. site study and plant relocation
23. business plans for new technology
24. industry market evaluation
25. pipeline corrosion study
26. environmental permit application

Omni Tech has conducted projects for chemical companies of all sizes; small, nonchemical industrial companies; engineering companies; plastics producers and fabricators; auto manufacturers; auto industry suppliers; metal platers; instrument manufacturers; independent oil producers; cogeneration producers; venture capital organizations; banks; local government units; health service organizations; legal firms (as expert witnesses); Native American organizations; sheltered workshops; and telephone companies.

We are a part of the local business environment, and we look forward to our company's being a permanent institution as new retirees join us and others carry on. Four important criteria for any consultant or company are

1. to be obsessed with clients and their requirements
2. to underpromise and overdeliver
3. to be professional in all respects—quality, timing, and technically
4. to recognize that hard work is necessary and that, while it reduces risk, there will always be risks

Our efforts are paying off. Revenues have doubled each year. Omni Tech is a recognized business in the community. New consulting associates are seeking us out. Clients are calling us. We are all working harder than before we retired. All participants are experiencing a productive and rewarding life after retirement.

Chapter 12

Robotic Servicing on the Space Station Freedom

Dale S. Schrumpf

Congress directed the National Air and Space Administration (NASA) to use the Space Station Freedom (S.S. Freedom) program to advance automation and robotics (A&R) technologies as an element of a national effort to increase U.S. productivity both in space and on the earth. The major benefits of using A&R on S.S. Freedom are presented. NASA proposed a series of dedicated robotic manipulators that would be resident in the subsystems of the Freedom Station. Transfer of chemical fluids is considered to be one of many tasks that could be performed by robotic manipulators. We will address how to pick and use consultants and how to become a consultant for robotic type tasks on S.S. Freedom.

Purpose

President Ronald Reagan stated on January 5, 1984, "We can follow our dreams to distant stars, living and working in space for peaceful economic and scientific gain. Tonight, I am directing NASA to develop a permanently manned space station and to do it within a decade". In response to this presidential mandate, NASA is planning to launch S.S. Freedom in 1996. In concurrence with a congressional mandate, NASA is focusing serious attention on the use of A&R in future space systems. Figure 1 shows an artist's concept of Lockheed Missiles & Space Company, Inc.'s version of NASA's S.S. Freedom.

Productivity in S.S. Freedom

NASA is conducting new research programs aimed at acquiring a better understanding of how A&R can work in partnership with people in complex, long-duration space-system missions. The important question that needs to be answered is how to tailor the relationship between robots and people using them to maximize productivity.

Purpose of Robotic Servicing

The goal of putting men into space has always been to determine their ability to live there productively and then to do useful work. Some of the important functions for which robots can increase a space station's capabilities and cost effectiveness are the following:

- minimizing the use of crew time
- performing servicing activities such as resupply and maintenance
- performing dangerous work materials involving hazardous or extravehicular activity
- performing housekeeping functions such as inspection of the spacecraft itself
- handling materials and performing other difficult functions for humans in the microgravity environment

Lockheed's participation in the Freedom Station program include:
- science
  — laboratory modules

Figure 1. An artist's concept of Lockheed Missiles & Space Company's version of NASA's S.S. Freedom.

- platforms
- payload pointing system
- vehicle systems
  - thermal
  - solar array
  - mechanical
  - truss
- system support
  - software support environment
- manned systems
  - extravehicular activity (EVA) systems and tools
  - habitat module
  - crew systems
  - airlock outfitting

S.S. Freedom will require many new advances in the field of space-worthy robots with autonomous capabilities. In some fields, engineers must create excellence that will lead the world.

Types of Robotic Systems

Teleoperator Systems. The main purpose in advancing A&R is to provide tools allowing humans to become more productive. The key to success is increased productivity in every space-station mission. By *automation*, we mean the technology by which control of physical processes or devices can be exercised according to established rules and, normally, without human intervention. By *robotics*, we mean the technology and devices (sensors, effectors, and computers) for performing, under human or automatic control, physical tasks requiring human abilities.

Teleoperation for the Freedom Station will be required to provide the capability for assembly, repair, and servicing of various elements. Teleoperation is the extension of a person's sensing and manipulating capabilities to a location remote from him. To productively perform these tasks, collision avoidance, dextrous end effectors, and some form of mobility of the teleoperators will be required[1].

S.S. Freedom will have more A&R than any other space vehicle. We anticipate a lower cost of operation if the system is optimally designed and ground operations are effectively used. Increased automation adds to the systems efficiency because reliability is enhanced from abnormal situations. Robots, teleoperations, and telepresence allow tasks that are unsuited to humans alone and that reduce the exposure of humans to hazardous situations, to be performed[2].

Telerobotic Systems. NASA is developing a flight telerobotic servicer (FTS) to aid in the assembly, maintenance, and servicing of the Freedom Station. The FTS is a multiple-arm dextrous-manipulation system that works with the remote manipulator system (RMS), the orbital maneuvering vehicle (OMV), and with free flyers. The FTS will be used in the assembly of Freedom. The term *telerobotic* describes a system that is a combination of teleoperator and robotic control. The human operator acts as a supervisor communicating to a computer information about tasks goals, constraints, plans, contingencies, assumptions, suggestions, and orders. The operator would be receiving information about accomplishments, difficulties, concerns, and, as requested, raw sensory data. One human computer-command station can supervise many telerobots.

For satellite builders and their customers, the servicing capability of the Freedom Station provides a facility and a service to lengthen greatly the useful life of the spacecraft. The crew divides their time between caring for the station systems and operating and maintaining payloads. Crew productivity aboard the space station is essential to mission success. By using A&R, the crew can devote more work time to payloads than in previous manned programs.

The A&R needed by the Freedom Station would provide crew productivity for the functions of maintenance, repair, servicing, and assembly for all elements of the station. Technology to implement most of these requirements is either currently available or will be available in time for the initial space station. Trend analysis, fault diagnosis, and subsystem statusing are common to subsystems such as electrical power, guidance, navigation, control, information management, environmental control, and life support[3].

The S.S. Freedom servicing bay provides contamination and thermal control, micrometeoroid protection, and utilities transfer capability for work on a large variety of satellites and space-station free flyers. The *RMS* is a symmetrical large manipulator arm comprised of seven modular rotary joints. Force-moment sensors with built-in electronics are incorporated at each extremity of the arm joint to provide operational load data. Each extremity of the arm is terminated with an end effector functioning as an interface mechanism with external systems. The arm is locatable so that it can operate either from the mobile remote servicer (MRS), or from fixed locations on the Freedom Station independent of the mobile service center (MSC). The FTS is designed to be a smart front end for the RMS.

Added capability is gained by providing limited artificial intelligence to the controller for the various systems. Telepresence with dextrous capability also provides added resources to increase productivity, reliability, and capability. *Telepresence* is the ideal of sensing sufficient information about the teleoperator and task, and communicating this to the human operators in a sufficiently natural way that the operators feels they are physically present at the remote site[4].

Autonomous Robots. Autonomous robots do not require supervision after they have been assigned a task to accomplish. Initial automated systems, such as for utilities and life-support systems, are stationary units. Advances in robotics will lead to autonomous machines that move about and perform tasks like a human[5,6].

Marketing Your Services

When a consultant asks me the best way to become a consultant for S.S. Freedom, I reply: "Be a consultant on a similar program". It is the very best way to get a consultant contract today. You also need to market your skills every week you are under contract to new companies. We usually hire consultants who are working for other companies. Everyone wants a consultant who is in high demand.

Another way to break into a new company is learning all you can about their problems and how you can help solve them. Much skill is needed to find the real problem before the interview, but you only have a short time to sell the company on your skills. You need only find the problems that the client really needs to solve. Therefore, learn all you can about the company and the program before you go on the interview.

Hiring a Consultant

The best consultants are those referred to you by a friend or associate. The only problem in hiring really good consultants is that they are usually so booked up that it's hard to get them to work for you. Therefore, you may need to resort to interviewing a number of consultants to find the right one. One clue is asking the consultant if he or she can work 40 hours a week for the next two years for your company. A really good consultant refuses to be dominated by one company. A good consultant simultaneously works for three or four companies.

A must in hiring consultants is asking for referrals and telephoning all of them. Ensure the applicants list their last contract and the reason they no longer are working for that client. It's also important that they be questioned by your top technical employee to see if they are expert in the areas they are claiming[7].

Conclusions

This chapter has highlighted the need for A&R technologies to increase productivity both in space and on the earth. Congress has directed NASA to advance A&R technologies on the S.S. Freedom Program. I presented the major benefits of using A&R on the Freedom Station. I discussed a series of robotic manipulators that will be used on the program. I also addressed the issue of how to become a consultant on the S.S. Freedom Program and how to choose a consultant.

References

1. *Robotics for Commercial Microelectronics Progress in Space: Workshop Proceedings,* NASA: Goddard Space Flight Center, Greenbelt, MD, December 1987.

2. Sheridan, Thomas B., Ed., *Human Factors in Automated and Robotic Space Systems: Proceedings of a Symposium,* National Research Council: Washington, DC, 1987.

3. Rodriguez, G., Ed., *Proceedings from the Workshop on Space Telerobotic,* Jet Propulsion Laboratory, California Institute of Technology: Pasadena, CA, vols. 1–3, 1987.

4. *First Annual Workshop on Space Operations Automation and Robotics (SOAR '87),* NASA, Johnson Space Center: Houston, TX, 1987.

5. Firschein, Oscar, et al., *Artificial Intelligence for Space Station Automation: Crew Safety, Productivity, Autonomy, Augmented Capability,* Noyes Publications: Park Ridge, NJ, 1986.

6. Staugaard, Andrew C., *Robotics and AI: An Introduction to Applied Machine Intelligence,* Prentice Hall: Englewood Cliffs, NJ, 1987.

7. Weinberg, Gerald M., *The Secrets of Consulting: A Guide to Giving and Getting Advice Successfully,* Dorset House: New York, 1985.

Chapter 13

Consultation in Sensory Evaluation

Gail Vance Civille

When discussing sensory evaluation and its importance to the chemical industry, we realize the necessity of understanding each product's sensory properties, such as flavor, oral texture, appearance, fragrance, aroma, skin-feel, and hand-feel. Consumer products companies sell sensory properties. Understanding and clarification of these properties are keys to unlocking a product's success or failure. Flavor and texture properties are of great importance to companies dealing with foods, beverages, oral personal care, and health care products. Fragrance and tactile properties are the key to marketing household products, personal care products, paper/fabric, and fabric care products.

Whether you are a chemist or a market analyst, consumer response is the bottom line to the liking, accepting, and preferring of products and perception of performance. Indeed, there is an effect of sensory properties on perceived performance and perceived efficacy. For example, if you add a certain fragrance to a lotion, a consumer may perceive that lotion to be more moisturizing than another. Or, if you introduce a harsh sounding cover to a baby diaper, the consumer may perceive the diaper as less absorbent and harsher to the baby's skin. Although the outer cover never touches the baby's skin and has nothing to do with the diaper's performance, the consumer transfers the sensory perception into a measure of performance. It is very important for all of us in industry to understand how the perception of the sensory properties of a product influences the perception of other properties, such as performance, quality, consistency, and shelf-life. Understanding the sensory profile enables the researcher to have a better sense of what attributes are driving and pushing a product's acceptance or rejection.

Sensory evaluation is a scientific discipline that evokes, measures, analyzes, and interprets human responses to materials as perceived through the senses. As sensory consultants, our approach is two phased, business and technical.

From the business side, we are attempting to provide valid and reliable product information for sound business decisions. Marketing studies need not be restricted to central location or home placement tests. Analytical sensory testing can be done in which trained individuals perceive small differences between products by describing the sensory attributes in great detail. Knowing the right questions to ask and the correct correlations to draw can lead to valuable information. We also advise the business community about new advances in research on perception and we develop new sensory methods. Interest in sensory related research can cover research on animal or human sensory systems. In addition, our sensory evaluation input provides guidance on sensory principles and techniques that influence project management.

On the technical side, sensory analysts and consultants design and execute sensory systems. In our work across several areas of the consumer products industry, we interface initially with research and development (R&D) to help move products through the R&D process to the marketing/market research phase. In quality control and quality assurance work, companies are interested in ensuring their products' sensory attributes are consistent. From a consumer product standpoint, 50% of what is perceived as quality by the consumer is innate positive characteristics and lack of defects. The other 50% relates to consistency. It behooves a manufacturer to measure this consistency of a product's flavor and texture before it is shipped. Our technical focus covers project objectives in product development, product maintenance, and product improvement and optimization.

In product maintenance, we are concerned with quality control, cost reduction, and shelf-life. There, the project objective may be to maintain the sensory appearance and fragrance of a household cleaner. Our sensory plan would include the identification of a control and the definition of the key sensory attributes of the control. We can establish a system to measure routinely these attributes and the effect of changes in them on consumer acceptance. Eventually, the objective is to find the simplest test or a benchmark to assure routine product maintenance.

For product improvement, one example might be to deliver higher peanut flavor for a snack product. Some possible development approaches are to use more peanuts, roast the peanuts darker, or use flavor enhancers. The sensory methodology includes attribute tests, a descriptive flavor spectrum panel, and consumer testing. Results may show that for some products a vanilla booster improves the flavor of a peanut creme. However, for a chocolate product, roasting the peanut darker improves the peanut flavor in relation to the chocolate, because the dark roasted flavors tend to compliment each other.

Sensory expertise can also involve patent infringement defense. In this case, the objective may be to prove a product does not infringe on a patent based on sensory properties defined in the patent. Methodology would include a descriptive spectrum panel of experts to show that the products differed through clear description of the type and strength of specific flavor, texture, and appearance attributes.

Sensory evaluation, therefore, is an important key to unlocking the mysteries to a product's success or failure. No matter what role the chemist has in industry—from product development to quality control—the chemist must understand the sensory properties of a product to address both business and technical aspects. One of the biggest challenges in industry today is quality measurement and quality assurance. Sensory evaluation can provide many of the answers to understanding consumer products.

Chapter 14

Chemical Information Consultants: Selection and Vitalization

Robert E. Maizell

Chemical information consultants offer one or more of three broad services: locating and obtaining copies of articles, patents, and other documents; searching online data bases; and offering opinion, critique, and guidance based on analysis of a variety of information sources (e.g., online, printed, and unpublished, including personal contacts) and on their own experience and know-how. A more specific idea of what chemical information consultants can do can be gleaned from the following examples of assignments to several consultants:

1. Our company has started to develop a new technology that we think can solve a major chemical industry problem. What methods are now used to solve this problem and how satisfactory are they in terms of performance and cost? What are the opportunities for new technologies in this field? Should we pursue our technology any further? What will it take to get our new technology qualified and commercialized?

2. One of our chemicals is used in a certain industry, but its long-term future in this use is uncertain because one of our

competitors now sells it for less. What other industries might use a product of this type, and how do we explore these potential new applications? Make the necessary field contacts for us, including contacts with potential new customers.

3. Identify and evaluate from a technical perspective the patent position of one of our competitors.

4. We have a demanding application for a new material of construction. Certain specific performance requirements must be met. What materials exist at the commercial or R&D stage in the United States or abroad that would meet our requirements?

5. Help us improve our report writing program, including the indexing aspects.

6. We plan to send a team to explore opportunities in a Third World country. Help this team learn the pertinent facts about the country and summarize the trends, with emphasis on technological opportunities.

7. Recommend and develop a current awareness program that will keep us up to date on one of our key organic chemical product lines. Train one of our staff to operate and modify the program as needed.

8. We have an idea for a new instrumental approach to the analysis of certain products. What is the state of the art? Describe and evaluate what exists worldwide and recommend directions for us.

9. Help us evaluate a potential investment opportunity from a technological perspective and recommend a course of action.

10. One of our products is being purchased in quantity by a large company, but we don't know the use. Can you identify the use so that we can pursue the opportunity more effectively?

11. Give our patent attorneys a second review of the prior art that will permit them to make a better decision on the patentability of a product.

12. Help us develop a technical library and train one of our people to use online data bases.

13. Certain products that we use regularly have been deemed environmentally unacceptable and will be phased out. What is the timetable for the phaseout? What replacements are being developed and when will they be available?
14. Give one of our people a guided tour of the U.S. Patent Office.

Professional Qualifications for Information Consultants

An information consultant to the chemical industry should have one or more degrees in chemistry and significant industrial experience in chemical information work. A person who has an excellent track record as manager of a chemical information center in a large company (or has equivalent experience) and is otherwise qualified as described in this chapter should be able to do a good job as an information consultant.

A good consultant is thoroughly familiar with the major online data bases and systems. The consultant also will have the skills and resources to go far beyond online tools, as necessary. The consultant understands and can tap all of the major and many of the minor printed sources, many of which are not available online and which often go back many years before online technology became available. This point is especially important for patent searches.

Good consultants can obtain information that is not available in print, such as research in progress. (They can do this through an extensive network of friends and associates in universities, government, and industry—and in other ways.) They also have know-how, personal files, and other resources that may obviate the need for extensive or expensive searches.

Services Offered

Many information consultants offer a wide range of services. For example, not only can they conduct patent, literature, and other searches, but they can organize the findings in concise, summarized form and present only those that are pertinent. They can also critique the results (findings) and explain what they mean to you.

The best consultants will recommend actions to be taken as a result of the findings and will help you follow up on the recommendations, including, for example, identifying and contacting potential customers for your new products. They may set up meetings with potential customers and others and accompany you on these visits. Additionally, consultants can advise on identification, evaluation, and commercialization of new products and processes.

A good consultant can help you plan and implement a solid current awareness program that will keep your finger on the pulse of any field of interest to you. Consultants can help you evaluate, organize, and plan your technical information center or library; plan internal abstracting and indexing; organize and train for report writing systems for laboratory personnel; do the searches required for filings with the Environmental Protection Agency or the Food and Drug Administration; and perform related activities. Helping you select and train your information center personnel is another service offered. (A technical information center as used here should not be confused with a computer center.)

The consultant can tell you how to obtain articles, patents, and other documents quickly and how to get them translated into English if necessary. Some consultants will also get any needed translation done. (The Chemical Abstracts Service Document Delivery Service is superb, but not all documents are included in its scope.)

The ultimate service an information consultant can perform is helping a company identify, understand, and profitably commercialize new technologies, both products and processes. This means that the good consultant must be creative. Good consultants bring to their clients elements of entrepreneurship, intrapreneurship, and the skill that I call "interpreteurship"—explaining what it all means.

Consultants know where to stop. They realize that exhaustive investigations are not always needed and understand the law of diminishing returns. They will not attempt to usurp the territories of patent attorneys, toxicologists, or physicians, but they may work with or refer you to associates who are qualified professionals in these areas.

Care and Feeding of an Information Consultant

Clients will obtain best results from consultants by following certain simple care-and-feeding principles. For example:

1. Before the work starts, define the key project objectives clearly. Make sure that the consultant understands exactly what you want, why, how, and when. This includes reports or presentations and an understanding of who will prepare them. Talk with the consultant enough so that the right job is done and the job is done right.

2. Tell the consultant everything needed to complete the assignment. (This assumes that a secrecy agreement is already in place.) Consultants work best when they know the full background.

3. Make one contact in your company responsible for the project, but designate other people in your company whom the consultant may talk with as necessary.

4. Before the work begins, agree with the consultant on compensation that is fair and competitive. You may want to provide for bonuses if certain targets are met. Also consider providing a piece of the action for certain assignments—for example, a percentage of sales or profits or of licensing revenues. Pay on a timely basis (typically within 30 days) for services that are satisfactorily delivered. Remember that most consultants have small operations and find it difficult to work with payment schedules that are strung out.

5. Provide prompt feedback. If the consultant does well (or does not do well), say so and explain why. Honest praise can yield the client a reservoir of both performance and good will that will pay significant dividends.

6. Hold out to the consultant the "carrot" of future assignments only if the chances that these will materialize are considerably better than even. Otherwise, consultants could make unwarranted investments of their own time and funds in anticipation of something that may never materialize.

When To Consider Hiring an Information Consultant

When should a chemical industry manager consider hiring an information consultant? The answers depend on whether a company is large or small, although, in both cases, consultants in any field offer the advantage of flexibility. They can be hired to do specific jobs only, although they should be subject to recall as needed.

Large Companies. Here I define large companies as those with sales of about $250 million per year or more. A company that size can frequently afford to hire one or more chemical information specialists full-time. Below this arbitrary level, it is more difficult to afford such persons full-time without giving them other assignments as well. Even the large company, with one or more full-time chemical information people, however, may wish to consider hiring an information consultant for various reasons:

1. At times of peak work load or during vacations when existing technical information staff cannot do the job. A key person may be unavailable for a number of reasons. Or there may be a special project that existing staff does not have the time to handle, such as compilation of a large number of Material Safety Data Sheets.

2. When the matter is especially critical—involving much money, for example, or a tight deadline. Any situation where the stakes are high is critical.

3. When an important technology is involved that is especially complex or new to the company, or new to chemistry, or very fast-moving.

4. When an independent or second opinion is needed—for example, a patentability search for a patent attorney.

5. If a fresh, objective outlook or perspective is desired.

6. When an information consultant can do the job more economically than can be done in house.

Small Companies. For smaller companies, which can't afford to hire a chemical information specialist full-time, or even part-time, the use of an information consultant is more compelling. Some reasons follow:

1. Information resources are often acquired randomly and may be scattered around the premises. The result is lost dollars, lost time, and lost productivity.

2. In most smaller companies, printed copies of important chemical information tools, such as *Chemical Abstracts, Kirk–Othmer Encyclopedia of Chemical Technology, Chemical Economics Handbook* (SRI), and the patent bulletins often are not readily available. Use of online data bases may not be effective because the laboratory chemists who do the searching have more important responsibilities and simply don't have the time to become and remain proficient with the online tools. Such deficiencies make it more difficult for smaller companies to understand the needs of larger companies, and hence to sell products to them, and even more difficult to compete with larger companies. The information chemist can train laboratory people to use online and other tools to best advantage and can help them keep up with changes in procedures or, alternatively, can conduct the searches, as appropriate.

3. Developing a new product or new use is usually difficult or impossible without studying the most recent and pertinent information. (However, companies that are perfectly comfortable doing what they have been doing for the past 5 or 10 years don't need an information consultant.)

Selecting and Hiring an Information Consultant

Sources of data on information consultants and brokers appear at the end of this chapter. In evaluating an information consultant, the chemical company executive will want to consider several factors:

1. Is the chemistry good between you and the consultant? Phone calls can help, but a personal meeting is much better in assessing this point.

2. Is the consultant's integrity beyond even a hint of reproach? Secrecy agreements are always recommended but must be coupled with an atmosphere of mutual trust between you and the consultant. Reference checks will help. Assure yourself that there are no conflicts of interest.

3. Are the services provided in person, by telephone, or by mail, and in what ratio? In person is optimum, especially during the initial and final discussions on a project, but not always

essential. The in-person meeting gives the consultant the opportunity to meet with all of your key staff, who should be involved and to study the project first-hand.

4. What special skills and experience does the consultant have? The more hands-on industrial experience, the better. Professional honors or election to professional society offices are often good indicators. How does the consultant keep professional skills up to date? Does the consultant belong to the ACS Division of Chemical Information? Lists of publications and patents should be examined, although some good consultants may have no papers or patents.

5. What information resources are readily available to the consultant? Access to online data bases alone is far from enough. Is there ready access to one or more major university libraries (and to other university resources); to selected industrial libraries under certain conditions; to the U.S. Patent Office Public Search Room or to U.S. Patent Depository Libraries? What about personal libraries?

6. What provision is there for handling rush assignments? Are any services provided during off-hours or weekends?

7. Can you get in touch with the consultant quickly and easily? Is there prompt response and coverage by answering machine or an answering service?

8. What physical resources are available to the consultant? For example, will reports be presented in neatly typed form? Can they be delivered by facsimile or via computer disk using a program compatible with yours?

9. What will the consultant's reports be like? If the results are raw, you may be wasting your money. You may want results that are screened and filtered for specific pertinence and critiqued for significance to the project at hand. You should expect full written reports with statement of objectives, scope, methods, results, interpretation of results, and recommendations—all clearly delineated and supported. The consultant should be a good communicator, able to explain and sell recommendations or ideas.

10. Is the consultant willing to pass the muster of a smaller, limited assignment for you before being assigned a large task?
11. What special connections, networks, or other affiliations does the consultant have? For example, some consultants have linkages to associates in independent analytical laboratories or small custom synthesis companies.

How To Control Consulting Costs

You can expect that consulting fees will be adjusted as assignments become longer term. In fact, most consultants probably would prefer to work on a project basis, and this is the best arrangement for most companies since it helps promote both continuity and responsiveness.

It is important to agree on cost estimates before the work is started. Estimates should include out-of-pocket costs, such as travel, for which the consultant is to be reimbursed. A written proposal from the consultant is desirable.

Be sure you understand and agree with the broad approach or strategies to be used.

You may want to break the assignment into phases so that you can authorize and manage the consultant's efforts on a controlled basis. If you have all you need by the completion of phase IV, you may not need phases V to X, but you and the consultant should agree on such an arrangement before the work is started.

You can further control the assignment by agreeing on time or other limits. For example, in certain kinds of patent and literature research you could limit initial efforts to identifying and evaluating patent documents in English that have been published or issued within the past 17 years. For a review of an emerging technology, coverage of the most recent five years may give you everything important you need to know.

Acknowledgments

Among those whose ideas and comments contributed to this paper are R. A. Hagstrom, Robert Blaker, W. V. Metanomski, Ben H. Weil, Edward P. Bartkus, Gabrielle S. Revesz, Paul Zurkowski, Betty

Unruh, Nancy Garman, and Helen Gordon. Their contributions are much appreciated.

Potential Sources of Information Consultants

Note: Listing in the following sources is not necessarily an endorsement. The usual interviews and credential and reference checks are recommended. Many good consultants are not listed in any of these sources.

1. Association of Consulting Chemists and Chemical Engineers, Inc., 310 Madison Avenue, New York, NY 10017. The ACCCE *Consulting Chemists Directory,* published periodically, is a source of consulting chemists of all types.
2. Burwell, H. *Directory of Fee-Based Information Services,* Burwell Enterprises, 3724 FM 1960 West, Houston, TX 77068. Annual. Ms. Burwell is president of a newly formed association of information brokers.
3. Maxfield, D. M. *Online Search Services Directory,* 2nd ed.; Gale Research: Detroit, MI, 1987. Includes more than 1700 libraries, information firms, and other organizations that perform online searches.
4. American Chemical Society Division of Chemical Information, 1155 16th Street, Washington, DC 20036. *Chemical Information Bulletin.* Issues of this publication usually contain several advertisements from chemical information consultants.
5. Unruh, B.; Cornog, M. *NFAIS Directory of Consultants and Contractors,* National Federation of Abstracting and Information Services, Philadelphia, PA, 1987. Some of the people listed have technical backgrounds and could perform chemical information studies.
6. Information Industry Association, 555 New Jersey Avenue, NW, Washington, DC 20001. Directory of members lists some companies that are qualified to act as information consultants.
7. Local schools of library or information science may have faculty who are qualified or may have recommendations.

References

The articles below are part of a series on information consultants and information brokers.

1. Warner, A. S. *Online* **1988,** *12,* 20–24.
2. Rugge, S. *Online* **1988,** *12,* 48–50.
3. Strizich, M. *Online* **1988,** *12,* 27–31.
4. Vickers, P.; Wootliff, V.; Warren, L.; Boldes, F. R.; Everett, J. H. *Online* **1988,** *12,* 42–51.

INDEX

Index

A

Activities, mid-sized consulting firm, 127
Advertising for consultants, 37
Advisory consultant, definition, 33
Age vs. experience, 15
Agencies, part-time professional employment
 brief description, 99
 list, 109
Agreements
 long-term, pros and cons, 24
 short-term, pros and cons, 23
American Association of Retired Persons (AARP), Volunteer Talent Bank, 102
American Chemical Society (ACS)
 Employment Services, agency for part-time employment, 99, 102
 file of retired chemists, source of associates for small consulting firm, 112
Areas of expertise, consulting firms, 5
Arthur D. Little, brief description of specialty, 5, 6t
Associates, mid-sized consulting firm, 121
Association of Consulting Chemists and Chemical Engineers (ACCCE)
 classifier by consultant's expertise in materials or products, 90–91
 classifier by consultant's expertise in process and equipment, 92–93
 classifier by consultant's function, 88–89
 description, 65
 membership, route to finding clients, 85
 services to clients, 94
 services to members, 85
 survey, 68–82
Association of Part-Time Professionals, agency specializing in part-time employment, 99
Automation, definition, 134
Automation and robotics, studied in space station Freedom, 132
Autonomous robots, for space station, 136

B

Barriers to accessing federal laboratories, 57
Barriers to effective use of consultants, 26
Basic research, use of individual consultants, 24
Battelle Memorial Institute, brief description of specialty, 5, 6t
Bayh–Dole Act, enabled technology transfer, 56
Benefits of consulting, for a professor, personal perspective, 44
Benefits to university, STERMPS program with industry, 51
Board of directors, forming, for small consulting firm, 112
Booz Allen & Hamilton, brief description of specialty, 5, 6t
Business problems, use of consultants for, 30
Business venture, mid-sized consulting firm, 126

C

CalSec Consultants, consulting firm, brief description, 98
Capitalization, 16
Career Placement Registry, agency specializing in part-time employment, 100
Cash-flow problems, use of consultants for, 30
Center for Environmental Information, volunteering opportunities requiring technical input, 104

157

Characteristics of good consultants, 14,
  22, 84
Chelan Associates
  consulting firm, brief description, 98
  formation, 111
Chem Systems Group, brief description
  of specialty, 5
Chemical company, mid-sized, need for
  consultants, 122
Chemical consultants
  advantages of using, 7–8
  disadvantages of using, 8–9
  types, 4
  See also Consultants
Chemical consulting
  definition, 3
  focus in the 1990s, 17
  forecast, 16
  volume in U.S., 3
Chemical consulting firms
  principal fields of specialization, 5
  specialized, 18
  varied and difficult to categorize, 65
Chemical Education for Public
  Understanding Seminars, volunteer
  opportunities for teaching, 105
Chemical Hazards Information Team,
  volunteering opportunities requiring
  technical input, 104
Chemical information, need for
  consultants, 143
Chemical information consultants
  professional qualifications, 145
  reasons to hire, 148
  selecting and hiring, 149
  services offered, 143, 145
  sources, 152
  successful relationship with clients, 146
Chemists Group, consulting firm, brief
  description, 99
Client concerns about consultants, Kodak
  survey results, 38
Client–consultant relationship, See
  Consultant–client relationship
Clients, how to find them, 85
Commodities Research Corporation, areas
  of expertise, 7

Compensation, determining employee's
  worth, 96
Complementary products marketed by
  consulting firm, 126
Compliance with regulations, use of
  consultants to ensure, 31
Concluding a consulting relationship, 27
CONDUX, consulting firm, brief
  description, 98
Confidentiality, barrier to consultant
  effectiveness, 26
Conflict of interest, barrier to consultant
  effectiveness, 26
Congressional testimony, need for
  consultants, 117
Consultant(s)
  advantages of using, 7–8
  as project manager, 33
  backgrounds, ACCCE survey, 68–74
  barriers to effective use, 26
  billing, ACCCE survey, 79
  chemical information, 143–152
  criteria for selection, 22, 23t, 33,
    121, 149
  directories, 37
  disadvantages of using, 8–9
  fees in ACCCE survey, 75–78
  fees in Kodak survey, 34
  hiring, 137
  need to define specialty, 84
  personal attributes, 14, 22, 84
  potential, evaluation, 37
  scope sheet, 86–87
  selection, 33
  sensory, business and technical
    functions, 140
  services offered, 83
  situations for their use, 30
  three fundamental categories, 83
  types, 4, 33
Consultant–client relationship
  chemical information consultants, 146
  concluding, 27
  facilitating, 39
  general considerations, 39
  joint participation, 17
  maximizing efficiency, 11

INDEX

Consultant–client relationship—
Continued
  preliminary considerations, 9–10, 25
Consultantships
  long-term, pros and cons, 24
  short-term, pros and cons, 23
Consulting
  advantages and disadvantages, 84
  as a career, advantages, 14–15
  as second career, 15
  for space station Freedom, 136
  general description, 83
Consulting firms
  and marketing of new and
    complementary products, 126
  as chemical consultants, 4–6, 18–19t
  expenses, 35
  general, with chemical practices, 6
  growth correlated with drop in
    retirement age, 95
  principal fields of specialization, 5
  specialized, 18
  typical, 98, 109
  varied and difficult to categorize, 65
Consulting income, small firm, 107
Consumer marketing research,
  examples, 6
Consumer response to products and
  performance, 139
Contacts, route to finding clients, 85
Contingency fees, consultants, 36
Contingent employment (part-time),
  growth, 101
Corporate officials, concerns about
  treatment of part-time and temporary
  employees, 102
Corporations, innovative work options,
  101
Cost
  consultants, controlling, 151
  consultants vs. employees, 32
  consultation, proposal options, 11
  industrial association with STERMPS
    program in university, 50
Council for International Exchange of
  Scholars, agency for volunteers with
  financial support, 103

Criteria for selection
  associates, mid-sized consulting
    firm, 121
  consultant, 22, 33
  consultants or companies, 128
  information consultant, 149

D

Daily billing rate, consultants, 34
Defining specialty, best approach to
  consulting, 84
Defining the project, mid-sized consulting
  firm, 124
Developing products, need for consultants,
  126
Dewitt & Company, expertise in
  petrochemicals, 7
Dollar amount, chemical consulting in
  U.S., 3
Du Pont, consulting services, 6

E

Effective use of consultants, hindrances, 26
Employee relations problems, need for
  consultants, 122
Employment, part-time
  advantages and disadvantages, 97–98
  trends, 101
Environmental evaluation of a plant, need
  for consultants, 122
Environmental regulations, relative to use
  of instruments, need for consultants,
  122
Evaluation
  potential consultant, 37
  value added by consultant, 25
Experience vs. age, 15
Expertise, use of consultants for, 30
Experts under contract, mid-sized
  consulting firm, 121

F

Federal laboratories
  annual research budget, 55
  barriers to industrial access, 57
  number, 55
Federal Laboratory Consortium for Technology Transfer (FLC)
  description, 57
  list of points of contact, 59–61
Federal regulations, consulting to help clients come into compliance, 105
Fees
  charged by consultants, 34
  contingency, consultants, 36
  determining employee's worth, 96
  incentive, consultants, 36
  performance, consultants, 36
  proposal options, 11
Fields of specialization, consulting firms, 5
Flight telerobotic servicer, for space station, 135
Food product, need for consultants, 117
Forty-Plus Club, agency for volunteers with financial support, 103
Freedom space station, purpose, 131
Functional consultant, definition, 33
Funding, immediate, use of consultants for, 31

G

General consulting firms
  with chemical practices, 6
  See also Consulting firms
GLM Telesis
  company formed by group of retired chemists, 98
  consulting firm, brief description, 99

H

Handling key personnel, use of consultants for, 31

Hazard Communication Standard, consulting to help clients come into compliance, 105
Hiring
  chemical consultant, 137
  information consultant, 149

I

Implementing organizational change, use of consultants for, 31
Incentive fees, consultants, 36
Individuals as chemical consultants, 4
Industrial access to federal laboratories, barriers to, 57
Industrial consulting services, types, 3
Industrial coordinator, role in STERMPS program with university, 50–51
Industrial expectations of university–industry relationships, 47
Industrial research contacts, benefits to a professor, 45
Industrial scientists, role in technical consulting relationship, 21
Industry–university relationships, *See* University–industry relationships
Informal interaction, ineffective barrier to consultant effectiveness, 26
Information for use in litigation, need for consultants, 117
Informational consultant, *See* Chemical information consultants
Instrument analysis, need for consultants, 122
Insurance problems, small consulting firm, 116
Inter-American Institute of Science, consulting firm, brief description, 99
International base, need, 16
International competitiveness, U.S. position, 53
International Executive Service Corps, agency for volunteers with financial support, 103

INDEX

International Voluntary Services, agency for volunteers with financial support, 103
Intertek Services Corporation, agency specializing in part-time employment, 100

J

Joint participation, consultant–client relationship, 17

K

Kline & Company
 brief description of specialty, 5
 growth since start-up, 16, 17
Kodak survey of consultants, 21–38
Kossoff & Associates, expertise in plastics, 7

L

Lab Support, agency specializing in part-time employment, 100
Labor officials, concerns about treatment of part-time and temporary employees, 102
Legislation
 enabling technology transfer, 56
 expected to encourage people to work longer, 95
Life Line, volunteering opportunities requiring technical input, 104
Little, Arthur D., brief description of specialty, 5, 6t
Long-term consulting relationships
 benefits, personal perspective, 43
 pros and cons, 24
Long-term research, use of individual consultants, 24

M

Management, individual consultants, 39
Management consulting, examples, 5, 6
Management's evaluation of needs, preliminary consulting agreement, 32
Manufacturer, small, need for consultants, 116
Manufacturers' consulting services, effect of expert systems, 7
Manufacturing plant, updating, need for consultants, 123
Market characteristics, for small consulting firm, 113
Market study consulting, multiclient surveys, 12
Marketing associates, mid-sized consulting firm, 121
Marketing methods, small consulting firm, 114
Marketing products, 126
Marketing research, need for consultants, 122, 124
Marketing study consulting, examples, 5, 6
Marketing your services
 as consultant for space station Freedom, 136
 as consultants, 84
McKinsey & Company, brief description of specialty, 5, 6t
Midwest Research Institute, brief description of specialty, 6
Multiclient surveys
 cost and examples, 13
 general description, 12

N

National Air and Space Administration, mandated to develop space station, 131
National Association of Temporary Services, list of employment services, 100
Negotiating business agreement, mid-sized consulting firm, 124

Network to facilitate technology transfer, 58

O

Objectivity
  as a benefit of consulting, 44
  use of consultants for, 30
Office building, need for consultants, 117
Omni Tech International, formation, 120
Operating practice, mid-sized consulting firm, 124
Operational consultant, definition, 33
Operational essentials for small consulting firm, 118
Organizational change, implementation, use of consultants for, 31
Organizational essentials for small consulting firm, 118
Organizational problems, use of consultants for, 31
Organizational steps in forming a consulting firm, 111

P

Part-time employment
  advantages and disadvantages, 97–98
  trends, 101
Part-time professional employment agencies, 99
Patent infringement defense, function of sensory consultant, 141
Patent Memorandum of 1983, 56
Peace Corps, agency for volunteers with financial support, 103
Perception, sensory, as measure of performance, 139–140
Performance fees, consultants, 36
Periodic reports, general description, 14
Personal attributes, consultants, 14, 22, 84
Personnel practices for older workers, 101
Petroleum company, need for consultants, 117
Plant location, need for consultants, 124
Political problems, use of consultants for, 31
Politicians, concerns about treatment of part-time and temporary employees, 102
Pooled fund arrangements, university–industry relationships, 49
Preliminary discussions, research consulting, 43
Pricing, proposal options, 11
Process consultant, definition, 33
Product improvement, function of sensory consultant, 141
Product maintenance, function of sensory consultant, 140
Product R&D, U.S. position, 54
Product venture, mid-sized consulting firm, 126
Product's success related to sensory properties, 139
Productivity aboard Freedom space station, 132, 135
Products, complementary, marketed by consulting firm, 126
Professional qualifications, information consultants, 145
Professors as consultants
  Kodak survey results, 21
  possibilities, 4, 44
Program director, university, role in starting STERMPS program with industry, 49
Project manager, consultant, 33
Proposals
  client consideration of, 11
  items for inclusion, 10
  price considerations, 11
Proprietary projects, as consulting assignments, 9

Q

Quality assurance program, need for consultants, 123

INDEX

R

Real estate transaction, need for consultants, 117
Reasons to become a consultant, for a professor, personal perspective, 44
Referral of consultants, 37
Referrals, route to finding clients, 85
Regional coordinator of Federal Laboratory Consortium for Technology Transfer, 58–61
Regulation compliance, use of consultants to ensure, 31
Remote manipulator system for space station, 136
Research
  basic, use of individual consultants, 24
  long-term, use of individual consultants, 24
Research and development expenditures in U.S., 55
Research budget, annual, federal laboratories, 55
Research consultants
  maximizing efficiency, personal perspective, 42
  preliminary sessions, 43
Research scientists, role in consultant–client relationship, 42
Resource acquisition, use of consultants for, 30
Resource Conservation and Recovery Act, consulting to help clients come into compliance, 105
Restructuring of industry, produced involuntary retirees, 96
Retirees, as chemical consultants, 15
Retirement
  early, attractive option, 96
  mandatory, banned by legislation, 95
Retirement age, continuing to drop, 95
Robotic servicing, purpose, 132
Robotic systems, types, 134
Robotics
  definition, 134
  studied in space station Freedom, 132
Robots, important functions in space, 132
RSVP (Retired Senior Volunteer Programs), agency for volunteers with financial support, 103

S

Salary, determining employee's worth, 96
Science and Technology Entry Program (STEP), volunteer opportunities for teaching, 105
Science Temps, agency specializing in part-time employment, 100
Scientific disciplines of individual consultants, Kodak survey results, 22
SCORE (Service Corps of Retired Executives), agency for volunteers with financial support, 103–104
Search strategies for qualified consultants, 36
Second Careers, agency specializing in part-time employment, 100
Selection
  associates, criteria, mid-sized consulting firm, 121
  best consultant for job, 33, 36
  information consultant, 149
Sensory consultants, business and technical functions, 140
Sensory evaluation
  description of discipline, 140
  importance to chemical industry, 139
Sensory methodology
  for patent infringement defense, 141
  for product improvement, 141
  for product maintenance, 140
Sensory perception as measure of performance, 139–140
Service Corps of Retired Executives (SCORE), agency for volunteers with financial support, 103–104
Short-term agreements, pros and cons, 23
Short-term problem-solving (STERMPS), university–industry program benefits to industry, 52

Short-term problem-solving (STERMPS), university–industry program—*Continued*
  benefits to university, 51
  costs to industry, 50
  general definition, 49
  how to set up, 49
  university–industry relationships, 49
Short-term problem resolution, benefit of university–industry relationship, 48
Software, marketing by consulting firm, 127
Soliciting industrial associates for STERMPS program with university, 50
Solo consultants
  pros and cons, 4
  reasons for entering business, 15
Southern Research Institute, brief description of specialty, 6
Southwest Research Institute, brief description of specialty, 6
Space station Freedom
  productivity increased by automation and robotics, 135
  purpose, 131
Specialty chemical plant, need for consultants, 124
Specialty groups, associates in mid-sized consulting firm, 125
Spill prevention program, need for consultants, 124
Springborn Testing Institute, brief description of specialty, 5, 6t
SRI International, brief description of specialty, 5, 6t
State regulations, consulting to help clients come into compliance, 105
STERMPS, *See* Short-term problem-solving
Stevenson–Wydler Act, enabled technology transfer, 56
Strategic Analysis (company), expertise in catalysts, 7
Student benefits, STERMPS program with industry, 51
Superfund Amendments Reauthorization Act, consulting to help clients come into compliance, 105
Supporting experts, reservoir, for small consulting firm, 112
Survey results
  ACCCE, 68–74
  individual consultants used by Kodak, 21
Syndicated services, general description and examples, 14

T

Team building
  for small consulting firm, 112
  mid-sized consulting firm, 121
Technical assistance
  temporary, 30
  use of consultants, 30
Technical feasibility studies, benefit of university–industry relationship, 48
Technical R&D consulting
  examples, 5, 6
  multiclient surveys, 12
Technology transfer
  and U.S. competitiveness, 56
  Federal Laboratory Consortium, 57
Technology Transfer Act of 1986
  implementation, 57
  network to facilitate, 57–59
Teleoperation, definition, 134
Teleoperator systems for space station, 134
Telepresence, definition, 136
Telerobotic systems for space station, 135
Testing and safety laboratories, examples, 5, 6
Townsend (Philip) Associates, expertise in plastics, 7
Training
  in handling hazardous materials, need for consultants, 123
  use of consultants, 32
Type of work
  chemical information consultants, 104, 116, 122, 143
  sensory consultants, 140

INDEX

U

University role in starting STERMPS program with industry, 49
University–industry relationships
 industrial expectations, 47
 types, 48
Updating of manufacturing plant, need for consultants, 123
U.S. competitiveness and technology transfer, 56

V

Value evaluation of consultant, 25
Visibility, route to finding clients, 85
VISTA (Volunteers in Service to America), agency for volunteers with financial support, 103
VITA (Volunteers in International Technical Assistance), agency for volunteers with financial support, 103

Volume, chemical consulting in U.S., 3
Volunteer opportunities
 possibilities, 102
 requiring technical input, 104
 teaching, 104
 with financial support
  list of agencies, 110
  programs, 103
Volunteer Talent Bank, AARP, 102

W

Waste disposal system, need for consultants, 123
Wastewater permit, evaluation, need for consultants, 124
Weighting of consultant selection criteria, Kodak survey results, 23
Worker Equity Initiative of AARP, main thrusts, 101
Wright Killen & Company, expertise in process technology, 7

Indexing: Janet S. Dodd and Robin Giroux
Production: Paula M. Befard and Margaret J. Brown
Acquisition: Robin Giroux

Printed and bound by Victor Graphics, Baltimore, MD

The paper used in this publication meets the minimum requirements of American National Standard for Information Sciences—Permanence of Paper for Printed Library Materials, ANSI Z39.48–1984. ∞